MANAGING UNCERTAINTY
IN EXPERT SYSTEMS

THE KLUWER INTERNATIONAL SERIES
IN ENGINEERING AND COMPUTER SCIENCE

MANAGING UNCERTAINTY
IN EXPERT SYSTEMS

by

Jerzy W. Grzymala-Busse
University of Kansas

KLUWER ACADEMIC PUBLISHERS
Boston/Dordrecht/London

Distributors for North America:
Kluwer Academic Publishers
101 Philip Drive
Assinippi Park
Norwell, Massachusetts 02061 USA

Distributors for all other countries:
Kluwer Academic Publishers Group
Distribution Centre
Post Office Box 322
3300 AH Dordrecht, THE NETHERLANDS

Library of Congress Cataloging-in-Publication Data

Grzymala-Busse, Jerzy W.
    Managing Uncertainty in Expert Systems / by Jerzy W. Grzymala-Busse.
        p.   cm. -- (The Kluwer International Series in Engineering and
    Computer Science ; SECS)
        Includes bibliographical references and index.
        ISBN: 0-7923-9169-1
        1. Expert Systems (Computer Science)   I. Title.   II. Series.
    QA76.76.E95G789   1991
    006.3'3--dc20

Printed on acid-free paper.

Printed in the United States of America

*To my mother Estera-Maria
and my late father Witold*

C O N T E N T S

LIST OF FIGURES

LIST OF TABLES

PREFACE

This book was written as a textbook for the course Fundamentals of Expert Systems at the University of Kansas. The course was offered for both undergraduate and graduate students, mostly pursuing a computer science major. They had different backgrounds and many had had no previous exposure to an artificial intelligence course. Thus, the main assumption of the book is that the reader is not familiar at all with artificial intelligence.

This book is different from other books on expert systems because

1. Anything redundant is skipped, yet precise and sufficient explanations are offered. Formalism is kept to a minimum,

2. It offers a unique feature—an extensive and deep coverage of techniques of reasoning under uncertainty, crucial for expert systems. Even though advantages and disadvantages of different techniques are shown, no bias is displayed toward any particular method. All are presented as objectively as possible. The reader should notice that the community of expert systems is divided into many camps, each promoting their favorite methods,

3. It is not tied to a software package. Although a course in expert systems would include a programming project, I decided not to describe any such package. Thus, the book is not biased toward any particular programming environment,

4. Important general features of expert systems are emphasized, instead of detailed characteristics of specific systems,

5. It provides many exercises illustrating ideas and helping the student learn the material. Most existing books on expert systems are not qualified as textbooks on expert systems because exercises are not included at all.

The book may be used as a

1. Textbook for advanced topics of expert systems, namely, uncertainty in expert systems, when the main emphasis is placed on Chapters 4 to 7,

2. Quick reference on techniques of managing uncertainty for anyone working in the area of expert systems,

3. Textbook for a course in expert systems, if an emphasis is placed on Chapters 1 to 3 and on a selection of material from Chapters 4 to 7. There is also the option of using an additional commercially available shell for a programming project. In assigning a programming project, the instructor may use any part of a great variety of books covering many subjects, such as car repair. Instructions for most of the "weekend mechanic" books are close stylistically to expert system rules.

Contents

Chapter 1 gives an introduction to the subject matter; it briefly presents basic concepts, history, and some perspectives of expert systems. Then it presents the architecture of an expert system and explains the stages of building an expert system. The concept of uncertainty in expert systems and the necessity of dealing with the phenomenon are then presented. The chapter ends with the description of taxonomy of expert systems.

Chapter 2 focuses on knowledge representation. Four basic ways to represent knowledge in expert systems are presented: first-order logic, production systems, semantic nets, and frames.

Chapter 3 contains material about knowledge acquisition. Among machine learning techniques, a method of rule learning from examples is explained in detail. Then problems of rule-base verification are discussed. In particular, both consistency and completeness of the rule base are presented.

Chapters 4, 5, and 6 present quantitative approaches, while Chapter 7 describes the qualitative approach to uncertainty in expert systems. Although many different techniques are analyzed, all are described assuming zero knowledge of the corresponding theory on the side of the reader.

Chapter 4 describes three techniques to make inference under uncertainty: Bayesian, belief networks, and certainty factors. All three techniques are characterized by the fact that uncertainty is described by a single number (e.g., a probability or certainty factor). This chapter begins with an explanation of useful material on probability theory. It then explains how the Bayesian approach is used in PROSPECTOR. Belief networks are outlined in the next section. The chapter ends with the description of the certainty factor method, used in MYCIN.

Chapter 5 presents two approaches to uncertainty in expert systems: Dempster-Shafer theory and a method used in INFERNO. In both approaches, uncertainty is described by a pair of numbers, lower and upper bounds on a numeric value representing uncertainty.

Chapter 6 introduces three techniques to deal with uncertainty: fuzzy set theory, incidence calculus, and rough set theory. In particular, fuzzy logic used in PRUF is explained in detail. All three techniques characterize uncertainty by sets.

Chapter 7 deals with nonnumeric approaches to dealing with uncertainty. It starts with a brief exposition of modal logic and then describes four different approaches representing nonmonotonicity: nonmonotonic and autoepistemic logic, default logic, and truth maintenance system. Reasoning based on plausible theory is explained. The chapter ends with a brief description of two heuristic qualitative methods, based on the concepts of endorsement and core, respectively.

Acknowledgments

First of all, I would like to thank my daughter Ania, who was the first reader of the manuscript and who edited it and processed it on a microcomputer. Her endless remarks and changes considerably improved the text. My son Witek helped with final preparation of the text. His help is deeply appreciated.

I very much appreciate the time my friend Dr. Zamir Bavel spent discussing with me the arcane details of publishing. Dr. Walter Sedelow provided many valuable suggestions. I am also extremely grateful to Dr. Chien-Chung Chan, who was my former teaching assistant in the course Fundamentals of Expert Systems in two academic years, 1987/88 and 1988/89, and who was also my Ph. D. student, for his many constructive remarks.

Many students helped in improving the readability of the book; among them, I am especially grateful to M. Grobe, J. R. Kilgore, and M. Thun, as well as to L. C. Auman, J. F. Fischer, T. Gibson, M. Moore, J. W. Phelps, and C. M. Yeh from my fall 1987 class, and to M. Brust, S.-P. Chong, S. Gharagozloo, and K. Look from my fall 1988 class in Fundamentals of Expert Systems.

I am especially grateful to my wife Dobroslawa and children Ania, Witek, and Jasiu for their trust and encouragement.

MANAGING UNCERTAINTY
IN EXPERT SYSTEMS

C H A P T E R

1

INTRODUCTION

The development of knowledge-based systems, or computer programs in which the domain knowledge is organized as a separated part, is a rapidly expanding field in the area of artificial intelligence. Recently, a more sophisticated version of knowledge-based systems, expert systems, has come into the limelight. Capable of assisting and even replacing human experts, expert systems are considered a major accomplishment in artificial intelligence.

The success of expert systems can be largely attributed to the wide scope of commercial, industrial, and scientific applications, ranging from Wall Street finances to Main Street doctors' offices.

Expert systems provide sound expertise in the form of diagnosis, instruction, prediction, advice, consultation, and so on. They may also be used as training tools for new personnel or to interpret data or to monitor observations. Expert systems are applicable in medicine, the military, science, engineering, and law (see Waterman 1986).

An expert system is a computer program capable of representing knowledge and reasoning about it in a narrow domain of expertise. Work on one of the prototypes of the expert system, DENDRAL, modeling the mass spectrometer, was begun in 1965. The project was directed by E. A. Feigenbaum, a professor of computer science, in cooperation with J. Lederberg, a professor of genetics. Although the knowledge of DENDRAL was encoded in the form of production rules, it still resembled ordinary programs. While DENDRAL can be thought of as a prototype of expert systems, the system MYCIN, which was used to diagnose and treat bacterial infections, and was initiated in 1972 by E. F. Shortliffe, is considered to be the actual grandfather of expert systems.

An expert system is a special case of a knowledge-based system as well; an expert system *is* a knowledge-based system dedicated to specific tasks requiring a lot of knowledge from a particular domain of expertise. The expertise of an ex-

1

pert system is permanent. As a result, expert systems may simulate expensive and scarce human experts.

Representing human experts in a particular problem area, an expert system can act much as a human expert would. Even though it is dealing with a large amount of knowledge, an expert system can update it easily. An expert system can justify and explain its reasoning and the solutions that come about through its reasoning process. Acting within its particular domain of expertise, an expert system can exceed human performance, but cannot solve problems that are impossible for humans to solve.

Domains with few, costly experts and with a theoretical foundation that is incomplete or poorly understood are well suited for expert systems. In particular, when data or the way human experts solve problems are not explicitly defined, expert systems may prove successful. If, on the other hand, the algorithm that human experts are following is deterministic and simple, the data are certain, and there is a surplus of minimum-wage experts, an expert system is not likely to be recommended. When the problems to be solved need creativity, imagination, or common sense, people have an advantage over expert system technology.

An expert system must exhibit accuracy and reliability, yet often it is forced to reason in the presence of uncertainty. Although the concept of reasoning under uncertainty is so essential in the expert system field, the nature of uncertainty is not well explained. There are many reasons why the use of classical logic in the reasoning of expert systems is inadequate and uncertainty must be taken into account. For example, uncertainty may be caused by the ambiguity of the terms used in the knowledge domain. Data, processed by a system, may be uncertain because they are incomplete, inconsistent, unreliable, or inaccurate. Inconsistency may be caused by different opinions of a few experts or even different opinions of the same expert in different moments of time. Unreliable data may result from the fact that information about the occurrence of an event is not reliable and is, for example, only likely or probable to some degree measured by a degree of belief, as in probability. Moreover, values of many kinds of data are measured with inaccurate equipment. Knowledge, stored in an expert system, may be uncertain as well. For example, rules may not be categorical. Even MYCIN used some kind of a mechanism for reasoning under uncertainty. Recently many different ways of managing uncertainty in expert systems have been developed, backed up by corresponding theories.

While ability to perform under uncertainty is a basic requirement for expert systems, many of the existing commercial shells, or expert systems with a removed knowledge base, do not have any mechanism for dealing with uncertainty. At the same time, the need to develop adequate means for reasoning under uncertainty is not only well recognized but has attracted a lot of attention from expert system developers recently. All these reasons justify the main emphasis of this book on the management of uncertainty in expert systems.

Up to now most expert systems were developed to find a solution from the given class of all possible solutions. Systems periodically monitoring required behavior and taking necessary actions belong to that class. Recently a new kind of expert system, designing objects or planning a sequence of actions satisfying given requirements, has been developed as well. Such systems may design very large-scale integration chips or plan an aircraft carrier mission. Constructing such systems is much more difficult.

This development is closely related to a newer kind of system, deep-knowledge-based systems, also called *second-generation expert systems*. As defined by L. Steels in (1986), second-generation expert systems combine heuristic reasoning, based on rules, with deep reasoning, based on a model of the problem domain. Using second-generation expert systems, the extra bonus is a property called *graceful degradation* of the system. This property, typical for human experts, is defined by giving reasonable (but not quite correct) answers for questions out of the domain of expertise. The first-generation expert systems fail in such a situation, while the second-generation expert systems, having access to a deeper model, are capable of an additional search for a solution.

In this chapter the main principles of expert systems are discussed, beginning with the description of architecture and basic components of expert systems. Building of an expert system and the concept of expert system tools are then explained. The most important idea of the book, uncertainty in expert systems, is briefly introduced. Lastly, taxonomy of expert systems is presented. The details are studied in the subsequent chapters.

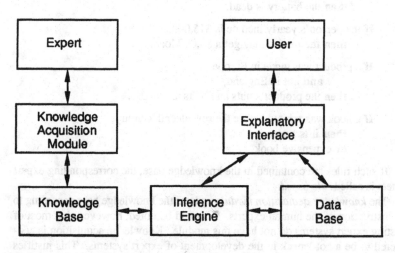

Figure 1.1 Architecture of an Expert System

1.1 Architecture of an Expert System

For the sake of simplicity, in this section we restrict the description to expert systems in which uncertainty is *not* taken into account. In the following sections of the chapter, this restriction will not apply.

Figure 1.1 illustrates the flow of information in a typical expert system.

A *data base,* known also as a *working memory,* or *short-term memory,* stores data for each specific task of the expert system. Some data are added to the data base through consultative dialogue between the expert system and the user, some by the inference of the expert system. After reaching a solution, the data base may be cleared in order to accept new data and start processing a new task. Typically, such data are *facts.* Examples of facts are

The engine won't start.

John is a man.

Ann's temperature is 98° F.

Bill has three cars.

A *knowledge base* (or *long-term memory*) contains general knowledge relevant to the problem domain. That knowledge is represented in an expert system in some way, as discussed in Chapter 3. At this point, to illustrate the concept, we restrict ourselves to an informal introduction of the most popular case of the knowledge base that contains *rules.* Rules have an If–Then format. Examples of rules follow.

If the ignition key is on,
    and the engine won't start,
    and the headlights do not work,
  then the battery is dead.

If the person's yearly income is $15,000,
  then the person may get a $3,000 loan.

If a product was made in Europe
    and not in England,
  then the product's nuts and bolts are metric.

If a book was printed before the seventeenth century,
  then it is a rare
  or expensive book.

If such rules are contained in the knowledge base, the corresponding expert system is called *rule-based.*

The *knowledge-acquisition module* updates the knowledge base according to the instruction of the human experts. It should be noted, however, that most of existing expert systems do not have this module. Knowledge acquisition is considered to be a bottleneck in the development of expert systems. This justifies the importance of algorithms for machine learning as a way to support the process of knowledge acquisition by a computer.

The part of an expert system that has the capability to infer is called an *inference engine*, or an *interpreter*. In the case of rule-based systems, the inference engine selects applicable rules from the knowledge base, identifying them by matching If clauses with facts from the data base, and applies one of the applicable rules to obtain new facts. By repeating this step, an inference chain is produced.

The following simple example illustrates the inference process. It shows how an inference engine uses a data base and knowledge base to determine whether a credit card application can be approved.

Suppose a data base consists of three facts:

    a. The person is asking for a credit card with a limit of $1,000,

    b. The person's yearly income is $15,000,

    c. The value of the person's house is $30,000,

and a knowledge base consists of three rules:

    1. **If** the person's yearly income is $15,000,
              and the value of the person's house is $30,000,
         then the person may get a loan of $3,000 or less.

    2. **If** the person may get a loan of $3,000 or less,
         then the person may get a credit card with a $1,000 limit.

    3. **If** the person asks for a credit card with a limit of $1,000,
              and the person may get a credit card with a limit of $1,000,
         then the person's application should be approved.

First, the inference engine compares the If clauses of each rule with each fact to identify the rules that are applicable. The If clause of rule 1 is matched, by facts b and c, and thus it may be applied. Rule 2, on the other hand, cannot be applied, since its If clause is not present in the data base.

After the application of rule 1, a new fact is added to the data base:

    d. The person may get a loan of $3,000 or less.

Then the If clauses of all rules are again compared with all facts. This time rules 1 and 2 may be applied. Selection of the appropriate rule will be discussed in Chapter 4, so for the time being let us assume that rule 2 is applied. That produces a new fact:

    e. The person may get a credit card with a limit of $1,000,

which is then added to the data base.

In the next step, the If clauses of all rules are compared with all facts. This time, rules 1, 2, and 3 may be applied. Rule 3 produces a new fact:

f. The person's application should be approved

and again, it is added to the data base. The last fact is a solution of the problem.

In the preceding example, inference is directed from data to solution. Such inference is called *forward chaining*.

An inference engine may also work backward, from a goal to data. The corresponding inference is called *backward chaining*. In some inference engines a mixed strategy of forward and backward chaining is applied.

An *explanatory interface* provides communication between the user and the expert system. It justifies the reasoning of the expert system and makes it possible for the user to ask how the expert system is reaching conclusions or why it is asking a specific question. Rule-based systems may answer by repeating the appropriate reasoning. The capability for explanatory reasoning should not be considered an option, since it may be crucial for an acceptance of the expert system in many applications.

In rule-based systems, in order to achieve the explanatory capability, rules may have the format If–Then–Because, where a Because clause is called an explanatory clause. This clause contains additional information given by an expert to justify the rule. These kinds of rules are used in COMPASS, for example, an expert system for telephone switching-system maintenance. An example of a rule with this format (not from COMPASS) is

> **If** the person may get a loan of $2,500,
>> **then** the person may get a credit card with a credit limit of $1,200,
>> **because** the statistical data of the bank show there is little risk.

1.2 Building Expert Systems

Building an expert system is a subject of *knowledge engineering*. The preliminary part of building an expert system, extracting knowledge from an expert and placing it in an expert system's knowledge base, is called *knowledge acquisition*. The person who interviews human experts is called a *knowledge engineer*. These experts include car mechanics, physicians, cooks, engineers, lawyers, and so on. The experts know the tricks and the "rules of thumb" in their area of expertise, making them especially useful. However, it should be clear that experts are not the only source of knowledge—books or journals are good ones as well.

1.2.1 Construction Stages

Building an expert system starts from the *problem analysis* (see Figure 1.2). A knowledge engineer, working with experts and managers at all levels, tries to identify the problem. The knowledge engineer must know how to describe the problem using natural language, as well as the goals and objectives for the expert system, the expectations of potential users, and their likely evaluation of the performance of an expert system. The very first decision that must be made is

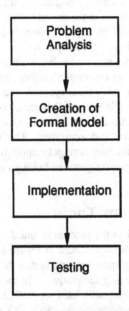

Figure 1.2 Stages of Expert System Construction

whether the problem is feasible, i.e., whether building an expert system is worthwhile in accordance with the criteria just discussed.

If the preliminary decision is to continue building an expert system, the next stage is the *creation of a formal model*. The main task is data elicitation, usually in verbal form, and analysis of data in order to recognize underlying knowledge. In this step essential concepts are defined, relations among them are established, procedures are analyzed, a hierarchy of subproblems of a main problem is recognized, and a hierarchy of strategies for solving the problems is explored.

The knowledge engineer must decide what form of knowledge representation will be used, what form of uncertainty is likely to appear, and a promising way to deal with it, as well as what tools might be needed to implement the program. On the basis of all of that, the knowledge engineer builds a formal model describing knowledge, the logical skeleton of an expert system. Decisions are made to select its detailed architecture, performance strategies for components of the expert system, and the relations between components and constraints. Then the final decision concerning the choice of tools to develop the expert system is made.

Next comes the *implementation*—creating a working program. The *testing* of a program follows. During the entire process of developing an expert system the knowledge engineer must closely cooperate with an expert. The role of an expert during this last stage cannot be neglected, since the expert is the one to evaluate and check the correctness of the expert system.

As a result, a *prototype* of an expert system is created. D. A. Waterman (1986) begins the classification of prototypes with a *demonstration prototype*, a small program with less than a hundred rules, working correctly for one or two cases. The main purpose is to encourage further funding of the project. The more developed prototypes evolve from systems that may fail on typical problems, through moderately reliable systems that are user-friendly and perform well in many cases, to systems that are well tested, reimplemented in an efficient programming language, fast, and accurate. The most developed system, a *commercial system*, is available, strangely enough, commercially. It takes approximately one to thirty person-years to build an expert system (Waterman, 1986).

1.2.2 Expert System Tools

Expert system tools are used in the process of building expert systems. Classification of expert system tools is presented in Figure 1.3.

One of the myths about expert systems is that they are written in either the LISP or PROLOG *programming language*. (On the other hand, it is true that many expert systems are developed in LISP or PROLOG.) LISP is based on the calculus of symbolic expressions, while PROLOG uses first-order predicate logic. LISP is preferred in the U.S., PROLOG in Europe and Japan. As a matter of fact, expert systems are implemented in all sorts of programming languages, including PASCAL, C, and BASIC.

Another possibility, aside from using programming languages, is to develop an expert system using a *shell*, or *knowledge engineering language*. A shell is usually based on some early expert system—it is an expert system with its domain knowledge removed. Commercially available shells are much more sophisticated. Commercial shells are equipped with some additional options, making their use much more convenient. The user may have a choice between ways

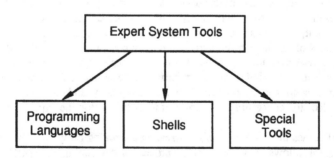

Figure 1.3 Expert System Tools

to represent knowledge, several strategies for inference, and so on. A shell may be used as a prototype, and the final product may be rewritten in an efficient programming language, such as C.

Special tools simplify tasks in expert system design, such as knowledge acquisition, knowledge-base design (checking rules for consistency, for example), explanatory interface design, and so on.

1.3 Uncertainty in Expert Systems

Expert systems achieve their high performance from extensive knowledge bases rather than from sophisticated algorithms. The know-how of experts, in most cases, follows from their practice and ability to guess well. Therefore, the knowledge of experts is not certain and expert systems often have to deal with uncertainty.

The program DENDRAL did not address uncertainty, but MYCIN did. Rules in MYCIN were qualified by a "certainty factor", for example,

> **If** A
> **then** B
> **with** certainty factor 0.8.

The concept of a certainty factor will be explained later.

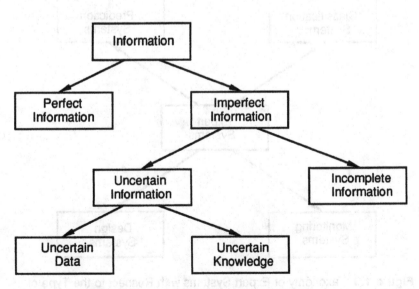

Figure 1.4 Classification of Information in Expert Systems

1.3.1 Sources of Uncertainty

Most of the following sources of uncertainty are quoted from Mamdani *et al.*
(1985):

- The *likelihood* of an event. Probability theory was developed especially
  to discuss this.

- Expert's *belief* in an event. One may have difficulty justifying such a
  belief.

- The *imprecision* of information. This may follow from real-world mea-
  surements of any kind. Another possibility is vague terminology, like
  "child abuse", "dangerous situation", and so on.

- An *exception* to a general rule. The general rule may be unknown (like
  in physics). Or it may be known but too complex to be efficiently implemented. A simpler rule is used, inexact but efficient.

- A *quantization* of a value. For example, the body temperature of a patient is quantized as being "normal", "a mild fever", "a fever", "a severe
  fever", and the limits are not precise.

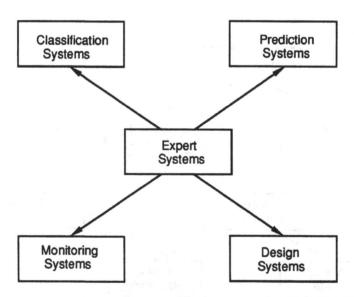

Figure 1.5 Taxonomy of Expert Systems with Respect to the Type of
Task

1.3.2 Inference under Uncertainty

In general, an inference chain is composed of elementary inference steps. Inference under uncertainty may involve uncertainty in data, or uncertainty in inference rules, or both. There are two basic approaches to uncertainty: *quantitative* and *qualitative*. In the former, the degree of uncertainty is represented by a measure (e.g., a single numerical value, like probability, or a set, like a fuzzy subset). The latter is represented by some nonclassical logics.

One of the problems is the validity of an inference composed of elementary inference steps, each taken under uncertainty. Another is the evaluation of the uncertainty of data obtained by such an inexact inference.

One classification of expert systems with respect to certainty of information was presented in Reichgelt and van Harmelen (1985). According to this method, expert systems may be forced to deal with different kinds of information, as illustrated by Figure 1.4. Information is perfect if it is true without qualifications; otherwise, it is imperfect. Imperfect information may be further characterized as incomplete or uncertain. In the case of incomplete information, not all instances of a problem are sufficiently described. An expert system may assume a default inference to hold true unless it receives new evidence contradicting the assumption.

1.4 Taxonomy of Expert Systems

There are many different classifications of expert systems. The two ways to classify expert systems described here are taken from Reichgelt and van Harmelen (1985).

1.4.1 By Task Type

The first taxonomy is based on the type of task that the expert system is performing. It is illustrated by Figure 1.5.

Classification systems (or *interpretation systems*) assign new classifications on the basis of low-level facts. For example, on the basis of the results of a patient's tests, the patient is classified as being ill, the disease is specified, and a treatment is prescribed. Classification systems use real data, given to an expert system either directly (by pick-ups or sensors) or indirectly as the result of some investigation. Typical classification systems uses include image analysis, speech understanding, medicine, and military uses.

Monitoring systems periodically check the current behavior of some system against the required behavior in order to detect malfunctions. Monitoring systems have been created for nuclear power plants, medicine, and manufacturing. In both classification and monitoring systems the solution space can be fully enumerated in advance.

Design systems construct complex objects that satisfy some conditions and constraints. If such a system constructs a sequence of actions, then the design system is called a *planning system*. Objects to be designed are, for example, cir-

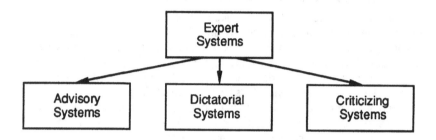

Figure 1.6 Taxonomy of Expert Systems Based on Interaction between User and Solution Status

cuit layouts and buildings. Planning systems are applied in a variety of areas, from automatic programming to military planning.

Prediction systems (or *simulation systems*) predict consequences of current situations. Prediction systems may incorporate a model to simulate real-world situations. Such systems may forecast weather or predict foreign currency exchange rates, crop yields, and so on.

Note that in design and prediction systems, the solution space is either impossible to enumerate or, when such full enumeration is possible, is not practical.

The preceding four types of expert systems may be used as primitives to build more complex systems.

1.4.2 By User Interaction and Solution Status

The second taxonomy of expert systems is based on the kind of interaction allowed between the user and the status accorded to the solutions offered by an expert system, as illustrated in Figure 1.6.

Advisory systems present solutions to the user, who is the final authority. The user may repeatedly reject solutions produced by the expert system. The system will try to look for other solutions after each rejection. *Dictatorial systems* produce the only correct solution (hopefully). Thus, a dictatorial expert system is treated as the final authority. *Criticizing systems* are confronted with a problem and a solution. The system analyzes and comments upon the solution.

C H A P T E R

2

KNOWLEDGE
REPRESENTATION

In this chapter, four representations of knowledge will be discussed: first-order logic, production systems, semantic nets, and frames. It should be noted, however, that there are other ways to represent knowledge.

The difference between "data" and "knowledge" is not quite distinct. The common agreement is that "data" represents "simpler" aspects of some universe, while "knowledge" describes the "sophisticated" aspects of the same universe. Knowledge is usually represented on a higher level than data (e.g., by rules, and its content may be qualified with respect to uncertainty).

2.1 First-Order Logic

From the viewpoint of knowledge representation, first-order logic provides a language for expressing knowledge and principles for manipulating expressions from this language. The simplest calculus of the first-order logic is propositional.

2.1.1 Propositional Calculus

Propositions are sentences that are either true or false. Thus,

"It is not raining now"

or

"The battery is dead or the terminals are corroded"

are propositions, the former *simple* and the latter *compound*, while

"Is it raining?"

or

Table 2.1 The Truth Table for Connectives

P	Q	$\neg P$	$P \vee Q$	$P \wedge Q$	$P \rightarrow Q$	$P \leftrightarrow Q$
T	T	F	T	T	T	T
T	F	F	T	F	F	F
F	T	T	T	F	T	F
F	F	T	F	F	T	T

"Change the battery"

are not propositions.

The truth values are denoted by T (true) and F (false). The *connectives*, also called *logical connectives*, contained in compound propositions, are "not", "or", "and", "if... then", and "if and only if" (denoted $\neg$, $\vee$, $\wedge$, $\rightarrow$, $\leftrightarrow$, and are called *negation, disjunction, conjunction, implication*, and *double implication*, respectively). The truth table for the logical connectives is presented here. Simple propositions are denoted P and Q.

A compound proposition whose truth value is T for every assignment of truth values to its components is a *tautology*. For example,

$$(P \rightarrow Q) \leftrightarrow (\neg P \vee Q)$$

is a tautology.

A compound proposition whose truth value is F for every assignment of truth values to its components is a *fallacy* (or a *contradiction*). An example of a fallacy is

$$(P \vee Q) \leftrightarrow (\neg P \wedge \neg Q).$$

Tautologies justify *inference rules*. One of the best-known inference rules is *modus ponens*,

$$\frac{P \wedge (P \rightarrow Q)}{Q}$$

or

from P and $P \rightarrow Q$, infer Q,

justified by the following tautology

$$(P \wedge (P \rightarrow Q)) \rightarrow Q.$$

Another well-known inference rule is *modus tollens*,

$$\frac{\neg Q \wedge (P \rightarrow Q)}{\neg P}$$

or

from not Q and $P \rightarrow Q$, infer not P.

The following proposition system of J. Lukasiewicz is an example of an axiom system of propositional calculus (Marciszewski, 1981):

$$(P \rightarrow Q) \rightarrow ((Q \rightarrow R) \rightarrow (P \rightarrow R)),$$

$$(\neg P \rightarrow P) \rightarrow P,$$

$$P \rightarrow (\neg P \rightarrow Q).$$

A *theorem* is an axiom or a proposition derived from axioms by rules of inference. Propositional calculus is characterized by the following properties:

Completeness. Any tautology P of the calculus can be derived using only the rules of inference. In other words, all tautologies of the propositional calculus can be proved.

Soundness. Only tautologies may be proved in the calculus.

Decidability. For any proposition P, there is an effective procedure to show, in a finite number of steps, whether P is a theorem or not.

2.1.2 Predicate Calculus

First-order predicate logic is a basis for PROLOG. In fact, the name PROLOG comes from PROgramming in LOGic. The expert system DADM (Deductively Augmented Data Management), designed to assist managers, has a knowledge base with logic representation (Rauch-Hindin, 1986). Predicate calculus is an extension of propositional calculus.

Expressions of predicate calculus may include the following symbols:

constants (Ann, Sue, 0, 1, 2),

variables (someone, x, y),

function symbols (f, likes, where $f(0) = f(1) = 0, f(2) = 1$, likes (Ann) = Sue, likes (Sue) = Ann),

predicate symbols or *relation symbols*, including the equality symbol (isnice, $\geq$, =),

connectives ($\neg$, $\vee$, $\wedge$, $\rightarrow$, $\leftrightarrow$), *quantifiers* ($\forall$, $\exists$),

parentheses and *commas* ((,), ,).

Symbol $\forall$ means *universal quantifier*, "for all", while $\exists$ means *existential quantifier*, "there exists". The expression to which a quantifier is applied is the *scope* of the quantifier. An occurrence of an individual variable x is *bound* if and only if it is either an occurrence ($\forall x$) or ($\exists x$) or within the scope of a quantifier ($\forall x$) or ($\exists x$). Any other occurrence of a variable is a *free* occurrence. For example, in the expression $(\forall x)(R(x) \wedge S(y))$, the scope of the quantifier is $R(x) \wedge S(y)$, and thus x is a bound variable in both of its occurrences. The variable y occurs free, since even if it is within the scope of a quantifier, the quantifier is not on y.

Thus, the following are expressions of predicate calculus:

Sue = likes (someone), i.e., someone likes Sue,

isnice (Ann),

Sue = likes (Ann) $\wedge$ Ann = likes (Sue),

$(\forall x) (1 \geq f(x))$,

$(\exists x) (x \geq f(2))$,

$(\forall x) (\exists y) ((1 \geq f(x)) \wedge (y \geq f(x)))$.

Predicate calculus is of first order if quantifiers $\forall$ and $\exists$ are taken over individuals (i.e., they are of the form ($\forall x$), ($\exists$someone)) but not over functions or predicates.

The set of all the preceding symbols is called the *alphabet* of the first-order predicate calculus. *Terms* are precisely those strings over the alphabet that may be obtained by finitely many applications of the following rules:

1. Every constant is a term,

2. Every variable is a term,

3. If $t_1, t_2,..., t_n$ are terms and f is an n-ary function symbol, then $f(t_1, t_2,..., t_n)$ is a term.

From the preceding examples, none is a term. Examples of terms are Ann, someone, likes(Ann).

Atomic formulas are expressions of the form

$$P(t_1, t_2,..., t_n),$$

where P is an n-ary predicate symbol, $t_1, t_2,..., t_n$ are terms, and $n \geq 1$. For example, isnice (Ann), $1 \geq 0, 2 = 2$ are atomic formulas.

The *well-formed formulas* are expressions that can be built from the atomic formulas by the use of finitely many times connective symbols and quantifier symbols.

The preceding definitions describe the *syntax* (grammar) of the first-order predicate calculus. Credit for the *semantics* for the first-order predicate calculus belongs to A. Tarski (1933). The concepts of truth and falsity in the first-order predicate calculus are furnished by the following definitions.

A *domain* is any nonempty set. Given a set X of well-formed formulas of the first-order predicate calculus, an *interpretation* of X is a domain D together with an assignment to each n-ary predicate symbol of an n-ary predicate on D, to each n-ary function symbol of an n-ary function on D, to each individual constant symbol of a fixed element of D, and to the equality symbol = the identity predicate = in D, where for $a, b \in D$, $a = b$ is true if and only if a and b are the same.

A well-formed formula of the first-order predicate calculus is *satisfiable in a domain* D if and only if there exists an interpretation with domain D and assignments of elements of D to the free occurrences of individual variables in the formula such that the resulting proposition is true. A well-formed formula is *valid in a domain* D if and only if for every interpretation with domain D and every assignment of elements of D to the free occurrences of individual variables in the formula the resulting proposition is true. A well-formed formula is *satisfiable* if and only if it is satisfiable in some domain, and it is *valid* if and only if it is valid in all domains. In propositional calculus valid formulas are tautologies.

Predicate calculus is characterized by completeness and soundness, but not by decidability.

2.2 Production Systems

Production systems were first proposed by E. Post in 1943, but their current form was introduced by A. Newell and H. A. Simon in 1972 for psychological modeling and by B. G. Buchanan and E. A. Feigenbaum in 1978 for expert systems.

A production system consists of

1. A *knowledge base*, also called a *rule base*, containing *production rules* (or *rules*, or *productions*),

2. A *data base*, containing *facts*,

3. A *rule interpreter*, also called a *rule application module*, to control the entire production system.

2.2.1 Production Rules

Production rules are units of knowledge of the form

If conditions

then actions.

The *condition part* of the production rule is also called the *IF part, premise, antecedent,* or *left-hand side* of the rule, while the *action part* of the rule is also called the *THEN part, conclusion, consequent, succedent,* or *right-hand side* of the rule. Actions are executed when conditions are true and the rule is fired.

The name *production rule* covers a whole spectrum of different concepts. The first way to classify production rules is with respect to the restrictions on logical connectives between conditions and actions. Usually production rules are of the following form:

$$C_1 \wedge C_2 \wedge \ldots \wedge C_m \rightarrow A_1 \vee A_2 \vee \ldots \vee A_n,$$

where $m, n \geq 1$ (see Subsection 2.3.2 and the definition of the *Kowalski form*).

An atomic formula $C_i, i = 1, 2,\ldots, m$, or $A_j, j = 1, 2,\ldots, n$, may be represented by a triple (entity, attribute, value); for example:

(person, weight, light),

(Jan, isnice, true),

or

(mycar, battery, weak).

In some cases (e.g., when the entity is known, is not essential, or is unique) triples may be replaced by pairs (attribute, value); for example:

(weight, small),

(isnice, true),

or

(battery, weak).

In the preceding examples attributes are unary functions or predicates, called *nominal descriptors* in Dietterich and Michalski (1983). Attributes that are k-ary functions or predicates for $k > 1$ are called *structured descriptors* in Dietterich and Michalski (1983). Examples of atomic formulas expressed by structured descriptors are

((node1, node2), arcs, 3),

((Lawrence, Topeka), distance, 20),

or

((Ann, Sue), likes, true).

In many expert systems production rules the form

$$C_1 \wedge C_2 \wedge \ldots \wedge C_m \rightarrow A_1 \vee A_2 \vee \ldots \vee A_n,$$

is reduced further to the list of the following forms

$$C_1 \wedge C_2 \wedge \ldots \wedge C_m \rightarrow A_1$$

or

$$C_1 \wedge C_2 \wedge \ldots \wedge C_m \rightarrow A_2$$

or

. . . .

or

$$C_1 \wedge C_2 \wedge \ldots \wedge C_m \rightarrow A_n.$$

The form $C_1 \wedge C_2 \wedge \ldots \wedge C_m \rightarrow A$ is the *Horn clause form*. Using this form, another production rule of the form

$$C_1 \vee C_2 \rightarrow A$$

should be substituted by the following two production rules:

$$C_1 \rightarrow A$$

and

$$C_2 \rightarrow A.$$

Another way to classify production rules is according to the kind of action. In general, an action is to *change* the content of the data base. Production rules in which actions are restricted exclusively to *add* facts to the data base are called *inference rules*.

In the case of production systems dealing with uncertainty, even more types of production rules are determined by the corresponding approaches to uncertainty. That will be discussed in the following chapters.

Note that production rules have *names* and may be equipped with additional features, not always following from uncertainty. One of them may be a *time tag*, to carry the information about the last time the production rule was used by the interpreter.

2.2.2 Data Base

A data base holds facts, usually in the form of triples (entity, attribute, value), where the entity is a constant; for example:

(Jan, weight, light)

or

(car# 42, battery, weak).

The content of the data base, in general, is changed cyclically by an interpreter. The facts may have time tags, so that the time of their insertion into the data base can be determined.

2.2.3 Rule Interpreter

The rule interpreter works iteratively in recognize-and-act cycles. In such a cycle, the interpreter first *matches* the condition part of the production rules against the facts in a data base, recognizing all *applicable rules*. Then it selects one of the applicable rules and *applies* the rule (*fires* or *executes* it). As a result, the action part of the production rule is inserted into the data base and the content of the data base is changed by the rule. Then the interpreter goes to the next recognize-and-act cycle. The interpreter stops its cycling when the problem is solved or a state is reached in which no rules are applicable.

2.2.3.1 Pattern Matching

Usually, the terms of conditions and actions of rules are written as triples (entity, attribute, value), where entities are variables, and facts are represented by similar triples, although of a different type, because the entities are constants. The problem of *pattern matching* arises, that is, matching triples of different types. For example:

> (person, yearly income, greater than $15,000)
> ∧ (person, value of house, greater than $30,000)
> → (person, loan to get, less than $3,000)

is a production rule, while

> (John, yearly income, greater than $15,000)

and

> (John, value of house, greater than $30,000)

are facts.

Before matching may be performed, the variable "person" must be assigned a constant value. The assignment of the constant "John" to variable "person" makes the first two patterns in the production rule identical to the corresponding facts. Thus, the firing of the production rule in forward chaining causes the new fact (John, loan to get, less than $3,000) to be added to the data base.

During a cycle performed by an interpreter, most of the time is spent on pattern matching. Thus it is important to find an efficient algorithm for pattern matching. The most popular one is the Rete match algorithm (Forgy, 1982). This algorithm takes advantage of pattern similarity and temporal redundancy. The former means that a rule is tested against the same contents of data base, and some matching can be done at the same time because many patterns are similar. The latter follows from the fact that the contents of the data base, although changed after each cycle, are modified only a little. Thus, for two consecutive cycles, most of the information necessary for pattern matching may be saved. The Rete matching algorithm is used in the rule-based language OPS5, a language used for programming expert systems.

2.2.3.2 Conflict Resolution

Recognition may be divided into *selection* and *conflict resolution*, where "selection" means the identification of all applicable rules, based on pattern matching, and "conflict resolution" means the choice of which rule to fire. Some approaches to conflict resolution are listed here—they may be used in combination.

- *The most specific rule.* Thus, if the facts in the data base are P and Q and the rules are $P \rightarrow R$ and $P \wedge Q \rightarrow S$, then both rules are applicable, and the second should be fired, because its condition part is more detailed,

- *The rule using the most recent facts.* Facts must have time tags,

- *Highest priority rule.* Rules must have assigned priorities,

- *The first rule.* Rules are linearly ordered and the least applicable rule is fired,

- *No rule is allowed to fire more than once on the basis of the same contents of the data base.* This eliminates firing the same rule all the time.

2.2.4 Forward Chaining

Forward chaining is also called *data-driven, bottom-up,* or *antecedent chaining.* During the selection time of each cycle, the interpreter is looking for applicable rules by matching condition parts of rules with the current contents of the data base. It is necessary to recognize when to stop applying rules. The condition to terminate the process is either when the goal is reached or when all possible facts are already inferred from the initial data base.

Consider the following trivial example of a production system that might be used in troubleshooting car problems. For simplicity, variables in rules are ignored in the example, so that pattern matching is not necessary. The rules are

R1. If the ignition key is on
      and the engine won't start,
   then the starting system (including the battery) is faulty.

R2. If the starting system (including the battery) is faulty
      and the headlights work,
   then the starter is faulty.

R3. If the starting system (including the battery) is faulty
      and the headlights do not work,
   then the battery is dead.

R4. If the voltage test of the ignition switch shows 1 to 6 volts,
   then the wiring between the ignition switch and the solenoid is OK.

R5. If the wiring between the ignition switch and the solenoid is OK,
   then replace the ignition switch.

Initially, the data base contains the following facts:

A. The ignition key is on,

B. The engine won't start,

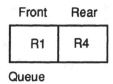

Queue

Figure 2.1 The Queue before the First Rule Firing

C. The headlights work,

D. The voltage test of the solenoid shows 1 to 6 volts.

In the example it is assumed that forward chaining is applied, and that a rule is applicable only if its condition part is true and if its action part adds a new fact to the data base. Moreover, rules are ordered according to their names (i.e., R1 precedes R2, R2 precedes R3, etc.) and they are scanned by the interpreter in that order. Applicable rules are successively inserted into a queue on a first-come, first-served basis. However, if a rule is already in the queue, then it is not inserted again. Conflict resolution is taken into account by firing the rule removed from the front of the queue. The goal is inferring all possible facts from the initial data base.

Initially, the applicable rules are R1 and R4, and they are inserted at the end of the queue, as illustrated by Figure 2.1.

Rule R1 is removed from the front of the queue and fired. Thus the new fact

E. The starting system (including the battery) is faulty

is added to the data base.

In the second cycle produced by the interpreter, rules are classified as follows:

- Rule R1 is no longer applicable, since its action part would add E to the data base, and it is already there.

- Rule R2 is applicable, and it is added to the end of the queue.

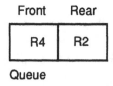

Queue

Figure 2.2 The Queue before the Second Rule Firing

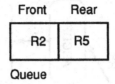

Queue

Figure 2.3 The Queue before the Third Rule Firing

- Rule R3 is not applicable; rule R4 is applicable, but it is already in the queue; rule R5 is not applicable.

Finally the queue contains two rules, R4 and R2, as illustrated by Figure 2.2. Rule R4 is removed from the front of the queue and fired, so that the new fact

F. The wiring between the ignition switch and the solenoid is OK

is added to the data base.

During the third cycle, after scanning the rules, rule R5 is inserted at the end of the queue (see Figure 2.3). Rule R2 is removed from the front of the queue and fired, producing the new fact

G. The starter is faulty,

which is then added to the data base.

As the interpreter discovers in the next cycle, no new rule may be inserted at the end of the queue, as the queue is represented by Figure 2.4. Rule R5 is removed from the front of the queue and fired, producing the new fact

H. Replace the ignition switch,

which is then added to the data base.

In the next cycle there are no applicable rules, nor are there any rules remaining in the queue from earlier cycles, so the interpreter halts the computation.

It can be seen that rules R1, R2, R3, R4, and R5 might be fired in different

Front = Rear

R5

Queue

Figure 2.4 The Queue before the Fourth Rule Firing

Managing Uncertainty in Expert Systems

Initial Content of Data Base

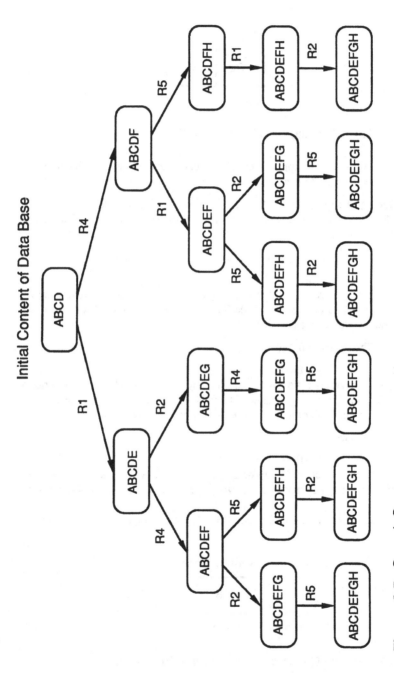

Figure 2.5   Search Space

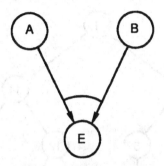

Figure 2.6 An Inference Network for Rule R1

sequences. The *search space* for the problem is presented in Figure 2.5, where nodes represent current contents of the data base and arcs depict applicable rules.

From Figure 2.5 it follows that with initial contents of the data base rules, R1 or R4 may fire. If the fired rule is R4, then rules R1 or R5 may fire, and so on.

The preceding sets of facts and rules may be represented by an *inference network*, a concept discussed in Duda *et al.* (1979). In the inference network, facts are viewed as nodes and rules as arcs. For example, rule R1, which says

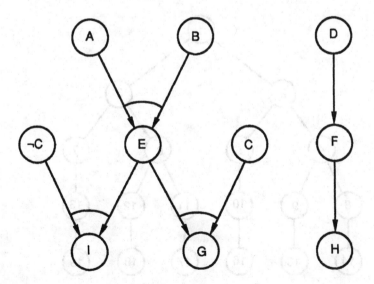

Figure 2.7 An Inference Network

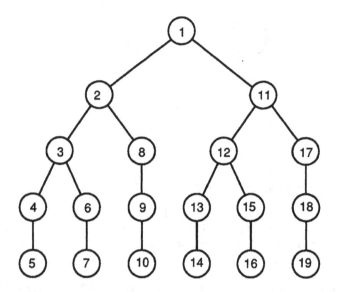

Figure 2.8 Depth-First Search

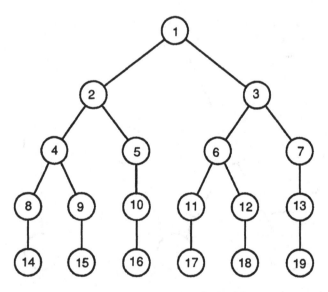

Figure 2.9 Breadth-First Search

If A and B then E

is presented in Figure 2.6.

The minor arc, connecting arrows from A to E and from B to E, indicates the **and** connective, which appears in rule R1 between facts A and B. The inference network illustrating the preceding example is presented in Figure 2.7, where I denotes the following fact:

The battery is dead.

Inference networks are helpful in solving problems associated with production systems. In some systems handling uncertainty, inference networks are used as useful models of knowledge propagation (see Section 4.2).

2.2.5   Depth-First and Breadth-First Search

The tree in Figure 2.5 may be traversed starting at the root and then following the leftmost path to a leaf, then starting from the last choice-point, again following the leftmost path to a leaf, and so on. This is depicted in Figure 2.8.

Another way of traversing the same tree is presented in Figure 2.9. Now the earliest choice-point is picked first, so that all nodes on a given level are visited from left to right before going on to the next level.

The main disadvantage of the depth-first search is that before the identification of the shortest path, many other long paths may be traced. The advantage is the simplicity of the implementation. Another advantage of the depth-first search is at the same time the disadvantage of the breadth-first search, namely, the breadth-first search needs more memory, since all nodes at a given level to the left and all nodes at the preceding level must be memorized. The advantage of the breadth-first method is that the first-found solution is always the shortest path.

2.2.6   Backward Chaining

In *backward chaining*, also called *goal-driven, top-down*, or *consequent chaining*, the production system establishes whether a goal is supported by the data base. For example, the goal is fact F. First, it should be checked whether F is in the data base. If so, F is supported by the data base. If not, but $\neg F$ is in the data base, the goal should be rejected. If neither F nor $\neg F$ is in the data base, applicable rules are determined. In **backward** chaining, applicable rules are recognized by matching action parts of rules with fact F. Let R be an applicable rule, selected by the interpreter. The condition part $C_1 \wedge C_2 \wedge ... \wedge C_m$ of rule R is now checked against the data base. If all $C_1, C_2, ..., C_m$, after the substitutions determined by matching, are in the data base, the solution is reached and F is true. Let's say that C is any of $C_1, C_2, ..., C_m$ (again, after corresponding substitutions). If $\neg C$ is present in the data base, then R cannot be used and another rule should be selected. If neither C nor $\neg C$ can be found in the data base, then C is a *subgoal* and the preceding procedure should start again the same way it is described for F. If no applicable rule exists and the truth of F is not established, the system may ask the user to provide additional facts or rules.

As an example, the same set of rules R1, R2, R3, R4, and R5 and initial content of the data base as in Section 2.2.4 will be considered. The action of rule R3

> The battery is dead

will be denoted by I.

The goal is $H \wedge I$. First, H will be considered. H is not in the data base. The only rule whose action part matches H is R5.

The condition part of R5 is F. F is not in the data base, so it is a subgoal. The only rule with the action part matching F is R4. The condition part of R4 is D, and it is in the data base, so F is supported, and hence H is supported. Next I must be checked. I is not in the base, so applicable rules are sought. The only such rule is R3. The condition part of R3 is $\neg\, C \wedge E$. C is in the data base, hence R3 can not be used. Because this is the only applicable rule to match I, I is not supported by the data base and the entire goal should be rejected.

Usually backward chaining is executed as depth-first search. Backward chaining is used in applications with a large amount of data (e.g. in medical diagnosis). If the goal is to infer a specific fact, forward chaining may waste time, inferring a lot of unnecessary facts. On the other hand, during backward chaining a very large tree of possibilities may be constructed.

Forward and backward chainings are two basic forms of inference in production systems. However, mixed strategies are frequently used in practice. For example, given facts and rules, forward chaining may be used initially and then backward chaining is applied to find other facts that support the same goal.

2.2.7 Metarules

Metarules are rules about rules. They may be *domain-specific*, like

> **If** the car does not start,
> > **then** first check the set of rules about the fuel system,

or *domain-free*, i.e., not related to the specific domain, like

> **If** the rules given by the owner's manual apply
> > **and** the rules given by the textbook apply,
> > **then** check first the rules given by the owner's manual.

Metarules reduce computation time by eliminating futile searching for a solution. The reduction may be achieved by *pruning* unnecessary paths in the search space, i.e., by ignoring such paths.

2.2.8 Forward and Backward Reasoning Versus Chaining

Forward and backward chaining were described previously. Both concepts are related to the way rules are activated by the interpreter.

On the other hand, the way *reasoning* is done depends on how the entire program is organized or what the problem-solving strategy is. If that strategy is bottom-up, the reasoning is forward. If it is top-down, the reasoning is backward (see Jackson 1986).

For example, XCON, an expert system designed to configure VAX computers, uses forward chaining and backward reasoning. The main goal, to configure a system, is divided into subgoals, to configure its components, and so on. Thus, the reasoning is backward.

2.2.9 Advantages and Disadvantages of Production Systems

The most obvious advantage of production systems is their modularity. Production rules are the independent pieces of knowledge, so that they may be easily added to or deleted from the knowledge base. The knowledge of experts is expressed in a natural way using rules, and then the rules may be easily understood by people not involved in expert system building.

The disadvantages follow from the inefficiency of big production systems with rules that are not organized in any kind of structure and from the fact that algorithms are difficult to follow. Still, rule-based expert systems are the most popular among all such systems.

2.3 Semantic Nets

Semantic nets and frames belong to a class of knowledge representations called "slot and filler" or "structured object". Semantic nets were first used by M. R. Quillian in 1968. At the same time, B. I. Raphael independently used the concept as a model of human memory. Semantic nets proved their usefulness in representation of natural language sentences.

A *semantic net* is a directed labeled graph, in which *nodes* represent entities, such as objects and situations, and *arcs* represent binary relations between enti-

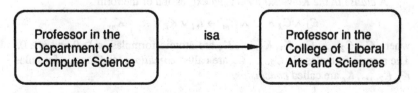

Figure 2.10 An *isa* Arc

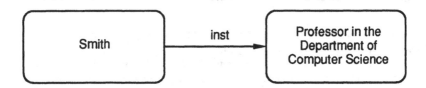

Figure 2.11 An *inst* Arc

ties. *Labels* on nodes and arcs represent their names.

2.3.1 Basic Properties

One of the key ideas in artificial intelligence is the *isa hierarchy*. The concept may be expressed using two predicates, *isa* ("is a") and *inst* ("is an instance of"). The former indicates that a class is a specific case of another class, while the latter says that a specific element belongs to a class. In both cases, objects of a specific nature *inherit properties* of objects of a more general nature.

For example, a class represented by a professor in the Department of Computer Science is a subclass of the class represented by a professor in the College of Liberal Arts and Sciences. On the other hand, Smith is an instance of a class represented by a professor in the Department of Computer Science. This is illustrated by Figures 2.10 and 2.11.

Semantic nets are natural representations for domains where reasoning is based on property inheritance. A simple example is presented in Figure 2.12. As shown, Smith uses CSNET, has a Ph.D. degree, and teaches CS 747.

A problem arises when we try to represent *n*-ary relations using semantic nets. For example, one wishes to express not only that Smith teaches CS 747, but also that he uses the textbook "AI" and that the location of his course is room 111 in Green Hall. To do that, an additional node is necessary, to represent the 4-ary relation "teaching", as illustrated by Figure 2.13.

2.3.2. Extended Semantic Nets

A version of semantic nets, *extended semantic nets*, was introduced in Deliyanni and Kowalski (1979). The main idea is based on "clausal form", introduced by R. A. Kowalski, also known as the "Kowalski clausal form" (Frost, 1986).

A *clause in the Kowalski form* is an expression of the form

$$C_1 \wedge C_2 \wedge ... \wedge C_m \rightarrow K_1 \vee K_2 \vee ... \vee K_n,$$

where $C_1, C_2, ..., C_m, K_1, K_2, ..., K_n$ are atomic formulas, $m \geq 0$ and $n \geq 0$. The atomic formulas $C_1, C_2, ..., C_m$ are called *conditions*, and atomic formulas $K_1, K_2, ..., K_n$ are called *conclusions*.

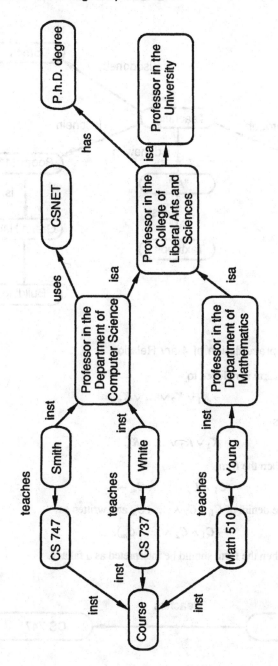

Figure 2.12　A Property Inheritance

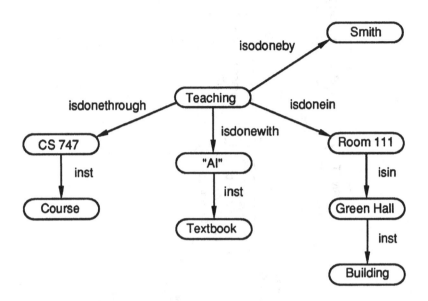

Figure 2.13 Representation of 4-ary Relation

For $m = 0$ the form is reduced to

$$\rightarrow K_1 \vee K_2 \vee \ldots \vee K_n$$

and is interpreted as

$$K_1 \vee K_2 \vee \ldots \vee K_n.$$

When $n = 0$, then the form

$$C_1 \wedge C_2 \wedge \ldots \wedge C_m \rightarrow$$

is interpreted as the denial of $C_1 \wedge C_2 \wedge \ldots \wedge C_m$ and written as

$$\neg(C_1 \wedge C_2 \wedge \ldots \wedge C_m).$$

If $m = n = 0$, then the form should be interpreted as a fallacy.

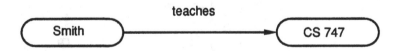

Figure 2.14 Conclusion Arc

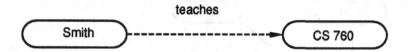

teaches

Smith - - - - - - - - - - - - - - -▶ CS 760

Figure 2.15 Condition Arc

In extended semantic nets, nodes represent terms, and arcs can represent binary relations. Two different types of arcs are used. Components of a condition are connected by the broken arrow, while components of a conclusion are connected by the ordinary arrow. Thus, the clause

→ Smith teaches CS 747

is represented by Figure 2.14, while

Smith teaches CS 760 →

meaning that Smith does not teach CS 760, is represented by Figure 2.15. Figure 2.16 illustrates the clause

Smith inst professor → (Smith inst male ∨ Smith inst female).

An example of the extended semantic net is presented in Figure 2.17. This net represents the following set of clauses:

→ Martin likes Robin,

→ Robin likes corn,

→ corn isa food,

Robin inst bird → Robin hasa beak,

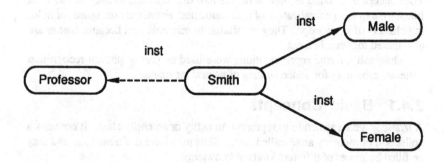

Figure 2.16 A Clause

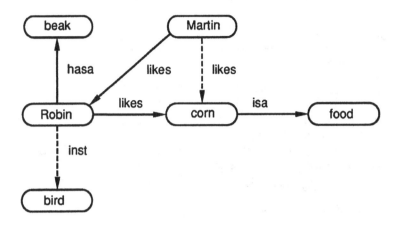

Figure 2.17 An Extended Semantic Net

Martin likes corn →.

2.3.3 Concluding Remarks

An example of the application of semantic nets to expert systems is PROSPECTOR, developed by Stanford Research Institute between 1974 and 1983. PROSPECTOR, a rule-based expert system (like KAS, a shell derived from it) uses a semantic net to organize production rules.

Inference in semantic nets is based on inheritance, but some other mechanisms are used as well, e.g., matching a fragment of the net with the entire net.

2.4 Frames

The concept of a frame as used here was introduced by M. Minsky in 1975. A frame system is a generalization of a semantic net. Frames are designed for holding clusters of knowledge. They are similar to semantic nets because frames are also linked together in a net.

Originally, frame representations were used as part of pattern-recognition systems, especially for understanding of natural language.

2.4.1 Basic Concepts

A *frame* is a data structure to represent an entity or an entity class. It contains a collection of memory areas called *slots*. Slots may have different sizes and may be filled by pieces of different kinds of knowledge.

Contents of slots may be categorized into *declarative* and *procedural*. Declarative content may be represented by *attributes* and their *values, descriptions*, or *graphical explanations, pointers* to other frames, *collections of rules*, and other frames.

Name Professor in the Department of Computer Science		
	Value:	Procedure:
Slot: Age		If-wrong
Condition	18 ≤ Age ≤70	
Slot: Ph.D. in		
Slot: Tenure		
Slot: Promotion rules		
Slot: Languages known	English	

Name Full Professor		
	Value:	Procedure:
Slot: Age		
Condition		
Slot: Ph.D. in		
Slot: Tenure	Yes	
Slot: Promotion rules		
Slot: Languages known	English	

Figure 2.18 Generic Frames

Name John Smith		Value:	Procedure:
Slot: Age		45	
Condition			
Slot: Ph.D. in		CS	
Slot: Tenure		Yes	
Slot: Promotion rules			
Slot: Languages known		English	

Figure 2.19 A Specific Frame

Some slots may contain, in addition, procedures that are triggered when the value of the attribute of that slot is changed. These procedures may have names like "If-added", "If-removed", "If-needed", and so on. They represent the procedural aspect of the slot content. Such procedures are called *demons*.

Frames are labeled by their names. When a frame represents a general concept, the frame is called *generic*. Frames containing specific information are said to be *specific* or *instantiated*. Examples of generic frames are given in Figure 2.18, and an example of a specific frame is presented in Figure 2.19.

Slot "languages known" in the frame "Professor in the Department of Computer Science" already has one of the possible values filled, i.e., "English". Such a value will occur in any instantiation of that frame and is called a *generic value*. Each frame from Figure 2.20 *inherits* that value.

Slot "tenure" in the frame "Full Professor" is filled by "Yes". It is expected that a full professor has tenure, so such a value is called a *default value*. If during the instantiation of that frame for a specific professor, it turned out that the status of "tenure" is "No", the default value will be changed.

In the frame "Professor in the Department of Computer Science", slot "Age" is furnished with a *condition* $18 \leq Age \leq 70$. An attempt to fill that slot with a value not satisfying the condition may trigger the attached procedure "If-wrong" and ask the user whether the value really is not an error in filling.

Frames may be linked together in taxonomical structures like semantic nets, using the same arcs "isa" and "inst", as shown in Figure 2.20.

Frame systems may be used to organize production rules, so the resulting knowledge representation is *hybrid*. Thus, rules are no longer unorganized, and

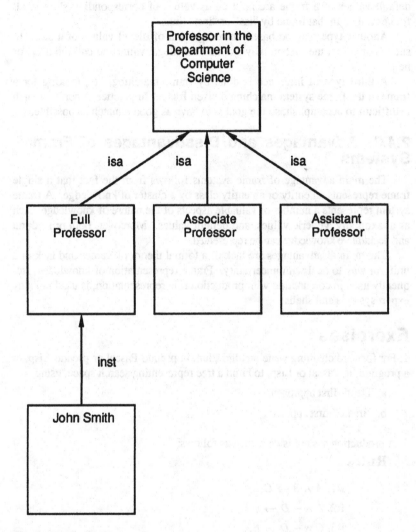

Figure 2.20 A Frame System

in one system advantages of both types of knowledge representation may be balanced.

2.4.2 Inference

Some types of inference can be made using frames. The first type of inference is based on structural properties of frames and taxonomical structures of a system. Some methods use inheritance, as in semantic nets. For example, generic and default values of a frame are inferred as values of corresponding slots in all frames linked to that frame by "isa" or "inst" links.

Another type may be based on recognition of illegal values of a slot. In such a situation, the system may exclude the illegal value and call the user for help.

A third type of inference is done by frame matching, i.e., looking for a frame in the frame system matching a given frame. In practice, a perfect match is difficult to accomplish, so the goal is to have as good a match as possible.

2.4.3 Advantages and Disadvantages of Frame Systems

The main advantage of frame systems follows from the fact that a single frame represents an entity or an entity class by a cluster of knowledge. A frame system represents a number of valuable aspects of the nature of knowledge, such as a taxonomy, generic values, and default values. Moreover, both procedural and declarative knowledge can be represented.

The main disadvantages are lack of a formal theory of frames and lack of a uniform way to deal with uncertainty. Frame representation of knowledge, frequently used in conjunction with production rule representation, is used in many expert systems and shells.

Exercises

1. For forward chaining write an algorithm, in pseudo-Pascal or pseudo-Lisp, or a program, in Pascal or Lisp, to build a tree representing search space, using

    a. Depth-first approach,

    b. Breadth-first approach.

2. A production system is described as follows:

    Rules

$$R1.\ A \wedge B \rightarrow C,$$
$$R2.\ A \wedge \neg D \rightarrow E,$$
$$R3.\ C \wedge \neg D \rightarrow E,$$
$$R4.\ C \wedge D \rightarrow F,$$
$$R5.\ E \wedge F \rightarrow L,$$
$$R6.\ E \wedge H \rightarrow \neg G,$$
$$R7.\ E \wedge \neg H \rightarrow G,$$
$$R8.\ I \rightarrow J,$$
$$R9.\ J \rightarrow K.$$

Conflict Resolution

  i. Rules are ordered according to their names.
  ii. The first applicable rule is selected.
  iii. During each session, each rule may be fired once.

a. Tell what the content of the data base is after forward-chaining session if the initial content of the data base is
$$\{A, B, \neg D, \neg H, I\},$$

b. As (a), but the initial content is
$$\{A, B, D, E, I\},$$

c. Tell whether the goal $\{L\}$ is supported by the following content of the data base
$$\{A, B, \neg D, E\}$$
(use backward chaining),

d. As (c), but the goal is $\{K, L\}$ and the content is
$$\{A, \neg D, \neg H, I\}.$$

3. A production system is described as follows:

Rules

R1. $A \wedge B \rightarrow C$,
R2. $\neg A \rightarrow H$,
R3. $B \wedge D \rightarrow \neg C$,
R4. $D \wedge E \rightarrow G$,
R5. $C \rightarrow I$,
R6. $\neg C \wedge G \rightarrow I$,
R7. $\neg C \wedge G \rightarrow H$,
R8. $\neg B \wedge \neg D \rightarrow J$,
R9. $I \rightarrow J$.

Conflict Resolution

  i. Rules are ordered according to their names.
  ii. The first applicable rule is selected.
  iii. During each session, each rule may be fired once.

    a. Tell whether the goal $\{H\}$ is supported by the following content of the data base:

$$\{A, \neg C, D, E\}$$

    (use backward chaining),

    b. As (a), but the goal is

$$\{J\} \; .$$

4. A production system is described as follows:

Rules

        R1. $A \wedge B \rightarrow D$,

        R2. $\neg A \rightarrow F$,

        R3. $\neg A \wedge B \rightarrow \neg E$,

        R4. $B \wedge C \rightarrow E$,

        R5. $D \rightarrow G$,

        R6. $D \wedge E \rightarrow I$,

        R7. $D \wedge \neg E \rightarrow H$,

        R8. $E \wedge F \rightarrow H$,

        R9. $\neg E \rightarrow G$.

Conflict Resolution

    i.   Rules are ordered according to their names.

    ii. The first applicable rule is selected.

    iii. During each session, each rule may be fired once.

Tell whether the goals

    a. $\{G\}$,

    b. $\{H\}$,

    c. $\{I\}$

are supported by the following content of data base:

$$\{\neg A, B, C\}.$$

Use backward chaining.

5. A production system is described as follows:

Rules

        R1. $A \wedge B \rightarrow E$,

R2. $A \wedge C \rightarrow F$,

R3. $\neg A \wedge B \rightarrow D$,

R4. $\neg A \wedge \neg B \rightarrow E$,

R5. $B \wedge C \rightarrow F$,

R6. $\neg B \wedge \neg C \rightarrow \neg F$,

R7. $D \wedge \neg F \rightarrow H$,

R8. $D \wedge F \rightarrow G$,

R9. $E \wedge F \rightarrow H$,

R10. $E \wedge \neg F \rightarrow G$.

Conflict Resolution

Rules are ordered according to their names. During each scanning of the list of rules by the interpreter, applicable rules are pushed into a stack successively. After scanning, a rule popped from the top of the stack is fired. During each session, any rule may be fired once.

a. Tell what the content of the data base is after the forward chaining session if the initial content of the data base is

$$\{A, B, C\},$$

b. Tell whether the goal $\{G\}$ is supported by the following content of the data base

$$\{\neg A, \neg B, \neg C\}$$

(use backward chaining),

c. As (b), but the goal is

$$\{H\}.$$

6. A production system is described as follows

Rules

R1. $A \wedge B \rightarrow \neg D$,

R2. $A \wedge B \rightarrow \neg E$,

R3. $A \wedge \neg B \rightarrow D$,

R4. $\neg A \wedge B \rightarrow D$,

R5. $\neg A \wedge C \rightarrow \neg E$,

R6. $\neg B \wedge \neg C \rightarrow E$,

R7. $E \wedge \neg D \rightarrow G$,

R8. $\neg E \wedge D \rightarrow G$,

R9. $D \wedge E \to H$.

Conflict Resolution

Rules are ordered according to their names. During each scanning of the list of rules by the interpreter, applicable rules are pushed into a stack, successively. After scanning, a rule popped from the top of the stack is fired. During each session, any rule may be fired once.

   a. Tell what the content of the data base is after the forward chaining session if the initial content of the data base is

$$\{A, B, C\},$$

   b. Tell whether the goal $\{G\}$ is supported by the following content of the data base:

$$\{A, \neg B, \neg C\}$$

(use backward chaining),

   c. As (b), but the goal is

$$\{H\}.$$

7. Represent by an extended semantic net the following set of clauses:

   $\to$ Sue likes Ann,

   Mary likes Ann $\to$,

   Ann likes math $\to$ Ann likes Bob $\vee$ Ann likes Sue,

   $\to$ Bob inst person,

   Sue likes Jan $\to$ Jan likes Sue.

C H A P T E R
3
KNOWLEDGE
ACQUISITION

The transfer of knowledge from some source into the knowledge base is called *knowledge acquisition* or *knowledge elicitation*. Knowledge acquisition involves applying a technique for elicitating data from the expert, interpreting data to recognize underlying knowledge, and constructing a knowledge representation model. Knowledge is acquired in chunks. After each such incremental increase of knowledge, an expert system's performance is also expected to improve.

The sources of knowledge are experts, books, personal experience, experimental data grouped in data bases, and so on. As a result of knowledge acquisition, the knowledge base is built. In the case of production rules, new rules are constructed and incorporated into the production system. In the case of semantic nets, new nodes are created and added onto the structure by adding links, while in the case of frame systems, some empty slots of existing frames may be filled. New frames may be created by instantiation of generic frames and then linked with the system. In all cases, the new units of knowledge may cause revision of an existing system (e.g., new rules may interfere with old ones, so that the entire system must be modified).

The addition of knowledge to the expert system may happen either *directly*, in the same units that are already in the knowledge base, or as the result of *induction* from examples, or *deduction* from knowledge other than that currently in the knowledge base. Both induction and deduction may be done by machine. Induction is the process of producing general principles from specific examples, while deduction means drawing specific conclusions from general principles.

In general, *knowledge acquisition techniques* may be categorized as *manual* and *computer-based* (see Figure 3.1). The most popular manual techniques of knowledge acquisition are based on a direct contact of a knowledge engineer with a domain expert without use of specialized computer-based tools. Manual tech-

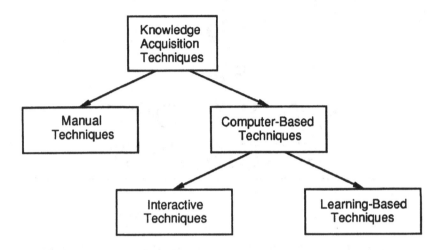

Figure 3.1 Knowledge Acquisition Techniques

niques of knowledge acquisition are time consuming and expensive. To simplify the process of knowledge acquisition, a variety of computer-based techniques have been developed. Computer-based techniques, used in knowledge acquisition, may be further categorized into *interactive* and *learning-based* (Boose, 1989). These techniques are discussed in the subsequent sections. A specific learning-based method, LEM, is presented in detail. The chapter ends with presentation of rule-base verification.

3.1   Manual and Interactive Techniques

Nowadays the basic technique of knowledge acquisition is still an interviewing of a domain expert by a knowledge engineer. This interview is done either directly, in an exclusive person-to-person way (i.e., manually) or with the help of some software package (i.e., by an interactive technique) (see Figure 3.2).

The two main ways to elicit knowledge are by interviewing the expert or by observing the expert. In the former technique, the knowledge engineer elicits the knowledge actively, by asking questions of the expert. The latter technique is based on the passive role of the knowledge engineer, who is observing the expert in action.

3.1.1   Interviewing

In a typical *interviewing session*, the knowledge engineer spends most of the time preparing the session, then analyzing the session, and the least time actually conducting the session. In the process of interviewing, a knowledge engi-

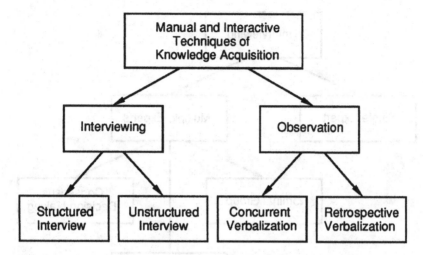

Figure 3.2 Manual and Interactive Techniques of Knowledge Acquisition

neer must take into account customer requirements for proper documentation of the knowledge base. In particular, customers wish to know where the knowledge base contents come from, for a variety of reasons, among them for the possibility of modification of the contents of the knowledge base in the future.

Usually sources of knowledge are organized in a specially designated *library* (McGraw and Harbison-Briggs, 1989). Concepts used in the domain are catalogued in a *knowledge dictionary*. A knowledge engineer acquires knowledge through a sequence of *interviewing sessions*. At the beginning of the process these sessions may be *unstructured* (e.g., the expert is giving a lecture), while the knowledge engineer is asking general questions. Later in the process of knowledge acquisition, interviews are mostly *structured* (i.e., most of the time the knowledge engineer is asking specific questions). Sessions may be audiotaped or videotaped. Some kind of record keeping must be used, in the form of making notes or filling out special forms, called *knowledge acquisition forms*. The knowledge acquisition form contains essential information, such as a list of session goals, description of elicited data in the form of facts, definitions, functions, relations, tables, taxonomies, diagrams, strategies, heuristics, rules derived from the session, as well as some official information (e.g., session time, number, location, name of the knowledge engineer, and name of the expert). As a result of interviewing, the knowledge engineer is producing rules of the form if-then, in plain English. Some kind of review process is used to verify acquired knowledge. In reviewing experts, other knowledge engineers or management may be used.

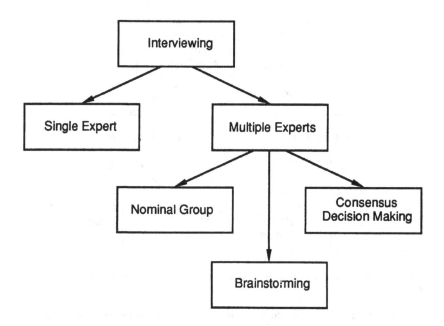

Figure 3.3 Interviewing

In a typical structured interview, the knowledge engineer should actively guide the expert through the session, follow the previously prepared agenda, call attention to the goal of the session, then summarize the session and verify results of the interview. At the same time, some pitfalls, as identified in many sources (e.g., Forsythe and Buchanan, 1989), should be avoided: The expert should be treated with respect; the knowledge engineer should keep a balance between being too domineering and not responsive enough. In general, the expert should have enough opportunity to express ideas; questions should be asked using the expert's language and should be brief. The process of interviewing is more art than science.

3.1.2 Observation

Sometimes experts cannot express their knowledge explicitly or may have difficulty doing so. In these cases it may be more fruitful to acquire knowledge by observation (i.e., by tracing the decision process of the expert). Two basic techniques used are *concurrent verbalization*, also called *thinking aloud* or *verbal protocol*, and *retrospective verbalization* (see Figure 3.2). The former technique is based on observation of the expert during the process of solving a problem, especially on recording the expert's thoughts during completion of the task. In the latter technique the knowledge engineer records the way the expert is solving the problem and then the expert explains the procedure to the knowledge engineer.

3.1.3 Multiple Experts

Usually, very complex domains of expertise require eliciting knowledge from many experts. In general, it is believed that a group of experts does better than an individual expert. The major advantage of multiple experts is the ability to detect and correct wrong answers. Special techniques of interviewing are used to deal with multiple experts (see Figure 3.3).

Multiple experts may be interviewed individually or in a team. Typically, in the beginning of the process of knowledge acquisition, the technique called *brainstorming* is used. At the beginning of the brainstorming session, its goal is briefly stated to the experts and then the experts express their ideas as quickly as they can, while their ideas are recorded and then analyzed. Later in the process of knowledge acquisition, another technique, called *nominal group*, may be used, where multiple-expert team members function independently. Yet another technique, *consensus decision making*, may be used. In this technique, a problem is presented to the multiple-expert team and, after some discussion, a solution is accomplished by voting.

3.1.4 Psychology-Based Techniques

Some knowledge-elicitation techniques are based on other disciplines (e.g., psychology, sociology, or anthropology). Most popular are psychology-based techniques (see Figure 3.4).

In a typical application of the *card-sort method*, concepts are typed on individual index cards and then are grouped together by the expert. The expert splits cards into smaller groups, then merges them into larger groups, using the think-aloud technique at the same time. *Multidimensional scaling* uses a similarity

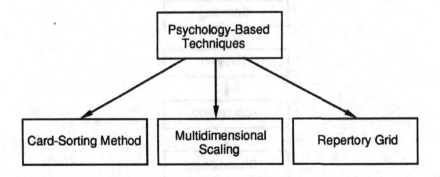

Figure 3.4 Psychology-Based Techniques

known to the expert to compare a set of objects. Using think-aloud technique, each object is compared with all others to make the similarity known to the knowledge engineer. Then a matrix of data is produced and the objects are scaled in a selected number of dimensions. A *repertory grid* is yet another psychology-based technique. In repertory grid, *elements*, used to define the area of expertise, are characterized by the expert according to *constructs*, whose values describe elements. For example, an element, a toy, is ranked by the expert according to construct "likable", on a scale from 0 to 9. Repertory grids may be further analyzed to find general patterns or principles. Multidimensional scaling and repertory grids are used to elicit knowledge in expert interviewing tools such as AQUINAS, KITTEN, KSSO, and PLANET (Boose, 1989).

3.1.5 Knowledge Acquisition under Uncertainty

A choice of a mechanism to handle uncertainty must be made early in the process of knowledge acquisition. The only help a knowledge engineer may get in making this decision is to be thoroughly familiar with as many theories as possible about dealing with uncertainty, and with their advantages and disadvantages.

Even in the case of building expert systems without any mechanism to deal

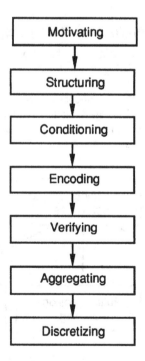

Figure 3.5 Probability Encoding

with uncertainty, the existing methodology does not provide any answer to the question of how to choose a representation of knowledge (e.g., whether to represent knowledge directly by production rules, by mixed representation of production rules organized by semantic nets, or by similar mixed representation of rules and frames).

There is little help available for deciding which tool should be selected to develop an expert system, even with respect to the preliminary decision of whether to use a general programming language or a knowledge engineering language, and if so, which commercially existing products to select. For obvious reasons, developers will try to sell their specific product.

There is no agreement between competing research groups, even when it comes to advantages and disadvantages of existing theories handling uncertainty. A process of knowledge acquisition is not objective. The choice of a way to deal with uncertainty is—in most cases—a subjective decision of a knowledge engineer. Once that decision is made, the rest of knowledge acquisition is biased. In particular, if the knowledge engineer had accepted a different decision, he would have asked the expert different questions.

One of the most detailed procedures for probability encoding based on structured interview was described in Merkhofer (1987). The main assumption is that values for subjective probabilities cannot be obtained directly from the expert because of accompanying biases. These biases are introduced by methods of reasoning by the individual expert or by a mechanism to elicit subjective probabilities from multiple experts. The procedure consists of seven steps (see Figure 3.5).

The first step is *motivating*, in which the knowledge engineer evaluates and removes biases from the reasoning of the expert. Two biases are particularly important: *management bias*, which may be exemplified by confusing management desires and goals with practice, and *expert bias*, meaning that the expert's knowledge—in the expert's mind—should not be uncertain. In the second step, *structuring*, essential variables whose uncertainty is to be assessed are selected and assumptions about variables are explored. The third step is *conditioning*. The knowledge engineer discusses with the expert knowledge related to the variables. Step four *quantifies* the uncertainty associated with variables. A variety of techniques may be used for that purpose. The expert may be asked to tell what values of the variable for fixed probabilities are, to tell what probabilities for fixed values of the variable are, or to tell both values of the variable and probabilities. Questions for probabilities may be asked directly or indirectly, by asking for best bets. The fifth step is *verifying*, to make sure that the expert believes in given probabilities, usually by transforming results of the former step and verifying if the expert is willing to bet money according to the rules. The next step, *aggregating*, is executed in the case of multiple-expert interviewing. Individual probability distributions are aggregated in this step. The last step is *discretizing*, or *quantization*. The range of values of the variable is selected from each interval, and then the probability, the actual value of which is taken from the interval, is assigned to that point.

3.2 Machine Learning

The question of learning was a subject of interest to the philosophers of ancient Greece. Today, one of the main goals of artificial intelligence is to equip computers with the ability to learn. The first attempts date back to the fifties.

The beginning of the current research on learning was initiated by P. H. Winston's 1970 Ph.D. dissertation (1975). The general assumption is that learning cannot be initiated without sufficient knowledge presented to the system. A learning system, INDUCE (Michalski and Chilausky, 1980), is well known as an example of a program outperforming human abilities in the quality of induced production rules.

This chapter is restricted to *empirical learning*, called also *similarity-based learning*, as opposed to more knowledge-oriented forms of learning such as *explanation-based learning*.

The main empirical machine-learning methods are rote learning, learning by being told, learning by analogy, and inductive learning (see Figure 3.6).

In *rote learning*, knowledge is directly supplied and memorized. No inference is required from the learner.

Learning by being told, also called *learning from instruction* or *advice taking*, is based on the adaptation of knowledge from the source to a form that may be accepted and utilized. The learner may perform some inference, but most of it

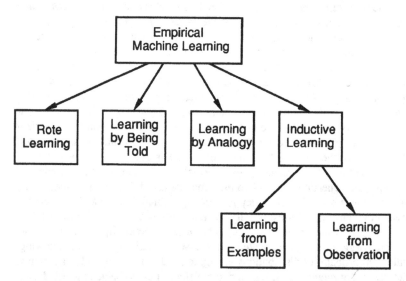

Figure 3.6 Machine Learning

is the responsibility of the teacher.

Learning by analogy is done by acquiring new knowledge that was helpful in doing a task in similar circumstances. This kind of learning requires more inference from the learner.

Inductive learning is categorized further into *learning from examples* and *learning from observation.*

In *learning from examples,* also called *concept acquisition,* a set of positive and negative examples is given, and the learner induces a higher-level concept description. This method was studied most intensively. Even more inference is required of the learner than in learning by analogy.

Learning from observation, or *descriptive generalization,* or *unsupervised learning* means creating new theories characterizing given facts. It asks the most of the learner, since it needs the most inference.

Among machine-learning methods, rule induction from examples seems to be practically the only one used in the expert system area (Waterman, 1986). Nevertheless, other aids exist to help a knowledge engineer. Some of them, namely checking rules from the knowledge base for consistency and completeness, will also be discussed in this chapter.

It should be noted that in inductive learning the learner may be exposed to only positive examples or to both positive and negative examples. One of the most important ways of categorizing methods of inductive learning is on the basis of possible description. Thus, the goal of learning may be to accomplish a *characteristic description,* in which a set of examples is described in such a way that it is distinguished from all other possible sets. This is the simplest form of inductive learning. Usually, the description is given as *maximally specific.* A well-known method of producing characteristic description has been developed in Winston (1975). In this work, the concept of near misses as negative examples and many others were introduced.

Another goal of learning is to come up with a *discriminant description.* In this case, a set of examples and a fixed family of other sets of examples are given. The task is to describe the set in such a way that it is distinguished from any set from the given family of sets. The description may be presented in a *minimal discriminant* form. Among the important contributions should be listed the algorithms based on AQ by R. S. Michalski and ID3 by J. Ross Quinlan (1983b).

Yet another goal of learning is presenting a *taxonomic description.* The problem is to classify a set of examples into subclasses satisfying some criteria. Recently, the most frequently used method is that of *conceptual clustering,* where sets of examples are classified into classes according to several concepts. METADENDRAL (see Buchanan and Feigenbaum, 1978) belongs to this area. Note that finding characteristic descriptions or discriminant descriptions constitutes learning from examples, while the determination of conceptual clustering belongs to methods of learning from observation.

Two approaches are used in inductive learning for control strategy. The first one is the *bottom-up* or *data-driven approach*, in which examples are described consecutively, in a number of steps, and their descriptions, more accurate after each step, are produced at the same time. The second is *top-down* or *model-driven*, where the set of possible descriptions is searched for a few optimal descriptions. The first approach is represented by the methods of Winston or Quinlan. The second approach is used in METADENDRAL.

Yet another taxonomy of learning systems follows from the type of acquired knowledge. The most popular way is to acquire production rules. An example of a learning system generating a *decision tree* is ID3. In a decision tree, nodes correspond to the attributes and the edges to attribute values. Other possibilities include learning semantic nets, frames, and classifications.

Most research activity in the area of inductive learning has been involved in creating methods of learning for complete input data, free of errors or conflicts. Such methods are described in this chapter. A method for learning rules under uncertainty is studied in Section 7.3.

3.3 Rule Learning from Examples

In rule learning from examples, large data bases or expert interview protocols are processed. As a result, production rules are delivered. In an interview protocol, examples or training data correspond to observations of what the expert did or what kind of actions were taken for specific situations. In such data bases, for many specific instances, values of *attributes* are determined for values of *decision variables*, briefly called *decisions*. Rule learning may be automated, thus provid-

Table 3.1 Patient Tests and Classifications

	Tests			Diseases		
	a_1	a_2 ...	a_l	b_1	b_2 ...	b_l
x_1						
x_2						
x_3						
$\vdots$						
x_n						

ing great help to a knowledge engineer. It should be emphasized (Buchanan *et al.*, 1983) that the efficiency of machine-learning programs is unimportant in comparison to potential benefits from improved performance of an expert system.

First, it should be clarified how "examples" are understood. To do so, the concept of a decision table is introduced. An algorithm for determining rules will be described later.

3.3.1 Decision Tables

The starting point for the approach presented here to an inductive algorithm is the *decision table*. Such a table may be a result of expert interview protocol, a retrieval of information from a data base, or just an observation protocol for some process.

For example, patients $x_1, x_2, ..., x_n$, all subjected to the same tests $a_1, a_2, ..., a_i$ (e.g., blood pressure, blood glucose, body temperature, and so on) and classified by physicians as being candidates for diseases $b_1, b_2, ..., b_j$ (e.g., diabetes mellitus type 1, diabetes mellitus type 2, diabetes insipidis, and so on), are presented in Table 3.1.

Another example is an industrial process (e.g., a sugar factory). At moments $t_1, t_2, ..., t_n$ of time, parameters $a_1, a_2, ..., a_i$ (e.g., temperature, pressure, concentration, in different places) and operator actions $b_1, b_2, ..., b_j$ (change of preset values of the steam flow, mixture flow, rotational speed, and so on) are recorded (see Table 3.2).

Tests or parameters $a_1, a_2, ..., a_i$ from the decision table are called *attributes*, while diseases or operator actions $b_1, b_2, ..., b_j$ are called *decisions*. An attribute

Table 3.2 Technological Process Parameters and Actions

	Process parameters			Operator actions		
	a_1	a_2 ...	a_l	b_1	b_2 ...	b_j
t_1						
t_2						
t_3						
⋮						
t_n						

or decision q may accept some *set of values*. That set is called the *domain* of q.

Any original set of values for any attribute may be *quantized* by experts, taking into account how they are making their decisions. For example, results of a glucose test may be classified as low, normal, high, very high level of sugar in the blood. Thus, four different levels are distinguished. Similarly, decisions may have a definite number of levels. A patient may be classified as being free from some disease, on a moderate level of it, on a severe level, or on a very severe level (the number of levels may differ).

Elements $x_1, x_2, ..., x_n$ from Table 3.1 or $t_1, t_2, ..., t_n$ from Table 3.2 are called, in general, *entities* of the table.

It should be noted that the concepts of a decision table and the concept of a data base, although similar, are not identical. A decision table may consist of duplicate rows, labeled by the two different entities x and y. This means that for any attribute or decision q, both entities, x and y, have identical values. Such a situation is not allowed in data bases. In decision tables it is quite normal. A lot of patients may have the same results of tests and be classified on the same level of disease. Similarly, process parameters and operator actions, recorded in different moments, may be equal, with the same accuracy, following from the quantization.

3.3.1.1   A Single Decision Table

Learning from examples will be illustrated first by a very simple decision table in Table 3.3. The attribute a may accept two values, L and H, where L stands

Table 3.3 A Decision Table

	Attributes		Decision
	a	*b*	*c*
x_1	L	0	u
x_2	H	0	u
x_3	L	1	w
x_4	H	1	v
x_5	L	2	w
x_6	H	2	w

for "low" and H for "high". Attribute b may have one of three values: 0, 1, and 2. Decision c represents three possible actions: u, v, and w. There are six entities $x_1, x_2,..., x_6$ in the table. Let x denote a variable that represents any entity. It is clear that the table is *complete*, i.e., all possible pairs of values for attribute pair (a, b) are included. Furthermore, there are no duplicate pairs of values of (a, b).

The entity x_1 may be described in the following way:

$$(x_1, a, L) \wedge (x_1, b, 0) \rightarrow (x_1, c, u),$$

that is,

> **if** for x_1 the value of a is L,
> **and** for x_1 the value of b is 0,
> **then** for x_1 the value of c is u.

It may be observed that the preceding rule is true not just for x_1 but for any x, because there is no other entity characterized by the same values for attributes a and b and characterized by another value for attribute c.

Thus, the rule describing x_1 may be written in the following form:

$$(x, a, L) \wedge (x, b, 0) \rightarrow (x, c, u).$$

The preceding technique is called *turning constants to variables* (Michalski, 1983). (Constant x_1 was turned to variable x.)

The only other entity for which attribute c has the value u is x_2, which may be described by

$$(x_2, a, H) \wedge (x_2, b, 0) \rightarrow (x_2, c, u),$$

or, after turning constant x_2 to variable x (it is allowed by the same justification as in the case of constant x_1), by

$$(x, a, H) \wedge (x, b, 0) \rightarrow (x, c, u).$$

There are only two possible values, L and H, for attribute a; therefore, the following two rules

$$(x, a, L) \wedge (x, b, 0) \rightarrow (x, c, u)$$
$$(x, a, H) \wedge (x, b, 0) \rightarrow (x, c, u)$$

may be replaced by one, simpler rule:

$$(x, b, 0) \rightarrow (x, c, u).$$

Either of the first two rules is said to be *subsumed* the third one. In other words, either of the first two rules has the same effect as the third one, but they contain additional restrictions.

This technique is called *dropping conditions* (Michalski, 1983).

Using similar argumentation, the next three rules may be constructed:

$$(x, a, L) \wedge (x, b, 1) \rightarrow (x, c, w),$$

Table 3.4 A Decision Table

	Attributes		Decision
	a	*b*	*c*
x_1	*L*	0	*u*
x_2	*H*	0	*u*
x_4	*H*	1	*v*
x_5	*L*	2	*w*
x_6	*H*	2	*w*

$$(x, b, 2) \rightarrow (x, c, w),$$
$$(x, a, H) \wedge (x, b, 1) \rightarrow (x, c, v).$$

Moreover, it may be observed that all four rules have the same variable x in each triple. For the sake of simplicity, we rewrite the rule in the form

$$(b, 0) \rightarrow (c, u),$$
$$(a, L) \wedge (b, 1) \rightarrow (c, w),$$
$$(b, 2) \rightarrow (c, w),$$
$$(a, H) \wedge (b, 1) \rightarrow (c, v),$$

so that x is ignored. In the next examples, the same simplification will be made—variables denoting entities will be ignored.

In the preceding example, rules were not systematically induced but rather guessed. Two algorithms for rule induction will be presented later on in this chapter. In rules built with the help of these two algorithms, decisions of induced rules depend on the smallest possible number of attributes.

The decision table from Table 3.3 is complete; hence all situations for attributes a and b are covered by the set of the last four rules. Therefore, there are no *missing rules*.

The concept of a missing rule is illustrated by Table 3.4.

Table 3.4 is a modified Table 3.3, namely, without the row labeled x_3. On the basis of Table 3.4, the following rules may be induced:

$$(b, 0) \rightarrow (c, u),$$

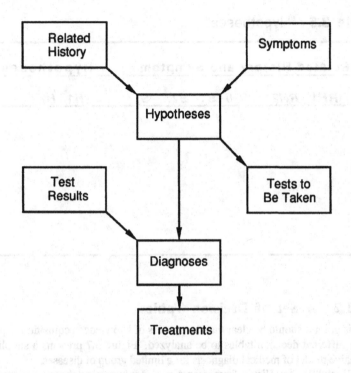

Figure 3.7 Medical Diagnosis

$$(b, 1) \rightarrow (c, v),$$

$$(b, 2) \rightarrow (c, w).$$

Comparing the preceding set of rules with the set of rules for Table 3.3, two observations can be made. First, rule

$$(a, H) \wedge (b, 1) \rightarrow (c, v)$$

obtained from Table 3.3, is simplified to

$$(b, 1) \rightarrow (c, v)$$

when it is derived from Table 3.4. Second, rule

$$(a, L) \wedge (b, 1) \rightarrow (c, w)$$

cannot be induced from Table 3.4 because of the lack of a row labeled x_3. That rule is missing.

Table 3.5 "Hypotheses"

	Related History and Symptoms						Hypotheses		
	RH1	*RH2... RHi*	*S1*	*S2... Sj*			*H1*	*H2...*	*Hk*
x_1									
x_2									
x_3									
$\vdots$									
x_n									

3.3.1.2   A Set of Decision Tables

At this point it should be clear that the process of knowledge acquisition requires many different decision tables to be analyzed. Figure 3.7 presents a simplified and naive model of medical diagnosis for a limited group of diseases.

Block "Related History" represents related information about a patient, like

Table 3.6 "Tests to Be Taken"

	Hypotheses			Tests to Be Taken		
	H1	*H2...*	*Hk*	*TT1*	*TT2...*	*TTl*
x_1						
x_2						
x_3						
$\vdots$						
x_n						

Table 3.7 "Diagnoses"

Hypotheses and Test Results						Diagnoses		
H1 H2... Hk	TR1	TR2...	TRm			D1	D2...	Dp
x_1								
x_2								
x_3								
⋮								
x_n								

personal data (like weight, height, age), recent diseases, surgeries, family history, and so on. Let's say that corresponding attributes are denoted $RH1$, $RH2,..., RHi$.

Block "Symptoms" comes from the patient's complaints to the physician. Corresponding attributes are $S1, S2,..., Sj$.

Block "Hypotheses" describes hypotheses denoted $H1, H2,..., Hk$. Each

Table 3.8 "Treatments"

Diagnoses			Treatments		
D1	D2...	Dp	T1	T2...	Tq
x_1					
x_2					
x_3					
⋮					
x_n					

Table 3.9 A Decision Table

	Attributes			Decision
	_a_	_b_	_c_	_d_
	terrain_familiarity	gasoline_level	distance	speed [m.p.h.]
x_1	poor	low	short	< 30
x_2	poor	low	short	< 30
x_3	good	low	medium	< 30
x_4	good	medium	short	30..50
x_5	poor	low	short	< 30
x_6	poor	high	long	> 50

such hypothesis corresponds to the level of severity of some hypothetical disease.

Block "Tests to Be Taken" depicts decisions to take or not to take medical tests $TT1$, $TT2$,..., TTl.

Block "Test Results" presents the results of the medical tests $TR1$, $TR2$,..., TRm.

Block "Diagnoses" shows diseases $D1$, $D2$,..., Dp. Each such disease is characterized by some level of severity.

Block "Treatments" characterizes what kind of therapy should be given. The corresponding attributes are $T1$, $T2$,..., Tq.

The four decision tables, presented in Tables 3.5, 3.6, 3.7, and 3.8, for patients x_1, x_2,..., x_n, will describe what is going on in the model. Each such table is a basis to induce rules.

3.3.2 Indiscernibility Relations and Partitions

An approach for rule induction from examples is presented here. The approach uses some of Z. Pawlak's concepts, presented in many papers (e.g., Pawlak, 1984). The presented algorithm was implemented as a Franz Lisp program LEM for VAX 11/780 (Dean and Grzymala-Busse, 1988).

An example of the decision table is presented in Table 3.9. Observed situations x_1, x_2,..., x_6 are described in terms of attributes: terrain_familiarity, gasoline_level, and distance, and driver's decision: speed of a car.

3.3.2.1　Indiscernibility Relations

Let Q denote the set of all attributes and decisions. In Table 3.9, $Q = \{a, b, c, d\}$. Let P be an arbitrary nonempty subset of Q.

Let U be the set of all entities. In the example $U = \{x_1, x_2, ..., x_6\}$. Let x and y be arbitrary entities. Entities x and y are said to be *indiscernible by* P, denoted

$$x \overbrace{}^{P} y,$$

if and only if x and y have the same value on all elements in P. Thus x and y are indiscernible by P if and only if the rows of the table, labeled by x and y and restricted to columns, labeled by elements from P, have, pairwise, the same values. In the example,

$$x_3 \overbrace{}_{\{a\}} x_4,$$

$$x_2 \overbrace{}_{\{b,d\}} x_3,$$

and

$$x_1 \overbrace{}_{Q} x_2.$$

3.3.2.2　Partitions

Obviously, the indiscernibility relation, associated with P, is an equivalence relation on U. As such, it induces a *partition of* U *generated by* P, denoted P^*. For simplicity, the partition P^* of U generated by P is also called a partition P^* of U, or yet simpler, if U is known, as partition P^*. As follows from its definition, partition P^* is the set of all equivalence classes (also called *blocks*) of the indiscernibility relation. Thus in the example

$$\{a\}^* = \{\{x_1, x_2, x_5, x_6\}, \{x_3, x_4\}\},$$

$$\{b\}^* = \{\{x_1, x_2, x_3, x_5\}, \{x_4\}, \{x_6\}\},$$

$$\{c\}^* = \{\{x_1, x_2, x_4, x_5\}, \{x_3\}, \{x_6\}\},$$

$$\{d\}^* = \{\{x_1, x_2, x_3, x_5\}, \{x_4\}, \{x_6\}\},$$

$$\{a, b\}^* = \{\{x_1, x_2, x_5\}, \{x_3\}, \{x_4\}, \{x_6\}\},$$

$$\{a, c\}^* = \{\{x_1, x_2, x_5\}, \{x_3\}, \{x_4\}, \{x_6\}\},$$

$$\{a, b, c\}^* = \{\{x_1, x_2, x_5\}, \{x_3\}, \{x_4\}, \{x_6\}\},$$

$$Q^* = \{\{x_1, x_2, x_5\}, \{x_3\}, \{x_4\}, \{x_6\}\}.$$

Partition $\{a\}^*$ has two blocks, $\{x_1, x_2, x_5, x_6\}$ and $\{x_3, x_4\}$.

Note that the concept of a partition, generated by P, makes specific values of attributes or decisions unimportant. These values are necessary to define a partition, but whether a domain of an attribute or decision is the set $\{0, 1, 2\}$ of integers, or the set $\{H, L\}$ of characters, or strings, does not make any difference because all partitions are defined on the same set U. Also, the fact that domains for different attributes or decisions are not disjoint is irrelevant.

3.3.3 Attribute Dependency and Rule Induction

In this section the concept of a covering, based on yet another concept, attribute dependency, is introduced. In the process of rule induction, redundant attributes in rules may be avoided, provided that these rules are constructed from coverings. This justifies the importance of concepts of attribute dependency and covering.

3.3.3.1 Attribute Dependency Inequality

Let P and R be nonempty subsets of set Q of all attributes and decisions. Set R is said to *depend on* set P if and only if

$$\widetilde{P} \subseteq \widetilde{R} \, .$$

The fact that R depends on P is denoted by $P \rightarrow R$. Note that $P \rightarrow R$ if and only if

$$P^* \leq R^*.$$

The preceding inequality is called *attribute dependency inequality*. Partition P^* is smaller than or equal to partition R^* if and only if for each block B of P^* there exists a block B' of R^* such that

$$B \subseteq B'.$$

The statement "set R depends on set P" may be characterized by the following: If a pair of entities cannot be distinguished by means of elements from P, then it cannot be distinguished by elements from R.

In the example, let $P = \{a, b\}$ and $R = \{d\}$. Then

$$P^* = \{\{x_1, x_2, x_5\}, \{x_3\}, \{x_4\}, \{x_6\}\} \leq R^* = \{\{x_1, x_2, x_3, x_5\}, \{x_4\}, \{x_6\}\},$$

so $\{d\}$ depends on $\{a, b\}$.

Similarly,

$$\{a, c\}^* \leq \{d\}^*,$$

so $\{d\}$ depends on $\{a, c\}$.

Moreover, $\{d\}$ does not depend on $\{a\}$ because partition $\{a\}^*$ is not smaller than or equal to partition $\{d\}^*$.

However, $\{d\}$ depends on $\{b\}$ because the dependency inequality is fulfilled:

$$\{b\}^* \leq \{d\}^*.$$

3.3.3.2 Equivalent Attribute Sets

It may happen that in the attribute dependency inequality, instead of weak inequality $\leq$, partitions $P*$ and $R*$ are related by equality. Formally speaking, in such a case R depends on P and P depends on R, or P and R are *equivalent*.

In our example,

$$\{b\}* = \{d\}*,$$

and

$$\{a, b\}* = \{a, c\}*,$$

so $\{b\}$ and $\{d\}$ are equivalent. Also, $\{a, b\}$ and $\{a, c\}$ are equivalent.

3.3.3.3 Coverings

Let R and S be nonempty subsets of the set Q of all attributes and decisions. A subset P of the set S is called a *covering of* R *in* S if and only if R depends on P and P is minimal in S. This is equivalent to the following definition: A subset P of S is a covering of R in S if and only if R depends on P and no proper subset P' of P exists such that R depends on P'.

In the example, the set Q has a covering $\{a, b\}$ in Q. Any two-element subset of Q, with the exception of $\{b, d\}$, is also a covering of Q in Q. The set $\{b, d\}$ has the following coverings in Q: $\{a, c\}$, $\{b\}$, and $\{d\}$. The same coverings as $\{b, d\}$ in Q have $\{b\}$ and $\{d\}$ in Q. Sets $\{a, b\}$, $\{a, c\}$, and $\{b, c\}$ are coverings of Q in $S = \{a, b, c\}$. Coverings of $\{d\}$ in $S = \{a, b, c\}$ are $\{a, c\}$ and $\{b\}$.

The concept of a covering was introduced to induce rules depending on as small a number of attributes as possible. In order to induce rules in which the THEN part consists of decisions from a set R and the IF part may consist of attributes from at most a set S of attributes, coverings of R in S should be used.

In the example the problem is to induce rules for $R = \{d\}$. Coverings of R in $S = \{a, b, c\}$ should be used. For any covering, for example, $\{a, c\}$, each entity from Table 3.9 will imply a rule for d (some of them are identical). All that should be done is to restrict attention to columns a, c, and d and write the rules. The fact that $\{a, c\}$ is a covering of $\{d\}$ in $\{a, b, c\}$ guarantees that—in general—both a and c are needed and that they are sufficient. Thus, the following rules are induced:

$$(a, poor) \wedge (c, short) \rightarrow (d, < 30),$$

$$(a, good) \wedge (c, medium) \rightarrow (d, < 30),$$

$$(a, good) \wedge (c, short) \rightarrow (d, 30..50),$$

$$(a, poor) \wedge (c, long) \rightarrow (d, > 50).$$

In this specific case, using dropping conditions, the preceding rules may be simplified to the following set:

$$(a, poor) \wedge (c, short) \rightarrow (d, < 30),$$

$$(c, \text{medium}) \rightarrow (d, < 30),$$

$$(a, \text{good}) \wedge (c, \text{short}) \rightarrow (d, 30..50),$$

$$(c, \text{long}) \rightarrow (d, > 50).$$

Although some rules are simplified, in general, both members, a and c, of a covering are necessary to express rules. It is also guaranteed by the concept of a covering.

Another possibility is the use of $\{b\}$ as a covering of $\{d\}$ in $\{a, b, c\}$. The corresponding rules are

$$(b, \text{low}) \rightarrow (d, < 30),$$

$$(b, \text{medium}) \rightarrow (d, 30..50),$$

$$(b, \text{high}) \rightarrow (d, > 50).$$

If the purpose is to induce rules for a decision q with the help of attributes from the set S, then coverings of $\{q\}$ in S should be computed first. Thus, the problem of finding coverings arises. The solution is the main part of the algorithm for rule learning. Subsections 3.3.5 and 3.3.6 are related to this problem.

3.3.4 Checking Attribute Dependency

The notation of 3.3.3.3 will be adopted here and in the next three subsections. In order to find a covering P of R in S,

1. P must be a subset of S,

2. Set R must depend on set P,

3. P must be minimal.

Condition (2) is true if and only if the attribute dependency inequality is fulfilled, i.e., if one of the following equivalent conditions (a), (b) is true:

a. $\widetilde{P} \subseteq \widetilde{R}$,

b. $P^* \leq R^*$.

The question is how to check (a) or (b) most efficiently. In order to check (a) for each set P, a new indiscernibility relation, associated with P, must be computed. Similarly, in order to check (b) for each P, a new partition, generated by P, must be determined. Both problems may be reduced for manipulation with single attributes. In case (a), the indiscernibility relation, associated with P, may be expressed in the following way

$$\widetilde{P} = \bigcap_{p \in P} \widetilde{\{p\}} .$$

Similarly, the partition of U, generated by P, may be presented by

$$P* = \prod_{p \in P} \widetilde{\{p\}^*} \; ,$$

where for partitions π and τ of U, $\pi \cdot \tau$ is a partition of U such that two entities, x and y, are in the same block of $\pi \cdot \tau$ if and only if x and y are in the same blocks for both partitions π and τ of U. Thus blocks of $\pi \cdot \tau$ are intersections of blocks of π and τ. For example,

$$\{a\}^* \cdot \{c\}^* = \{\{x_1, x_2, x_5, x_6\}, \{x_3, x_4\}\} \cdot \{\{x_1, x_2, x_4, x_5\}, \{x_3\}, \{x_6\}\}$$

$$= \{\{x_1, x_2, x_5\}, \{x_3\}, \{x_4\}, \{x_6\}\}.$$

In the example, let us say that $S = \{a, b, c\}$ and $R = \{d\}$. Then $\{b\}$ is a covering of $\{d\}$ in $\{a, b, c\}$ because

$$\{b\}^* \le \{d\}^*.$$

Another covering of $\{d\}$ in $\{a, b, c\}$ is $\{a, c\}$ (because $\{a\}^* \cdot \{c\}^* \le \{d\}^*$ and neither $\{a\}^* \not\le \{d\}^*$ nor $\{c\}^* \not\le \{d\}^*$). Sets $\{a, b\}$ and $\{b, c\}$, although $\{a\}^* \cdot \{b\}^* \le \{d\}^*$ and $\{b\}^* \cdot \{c\}^* \le \{d\}^*$, are not coverings of $\{d\}$ in $\{a, b, c\}$ because they are not minimal.

3.3.5 An Algorithm for Finding the Set of All Coverings

The aim of the algorithm is to find the set C of all coverings of R in S. The cardinality of the set X is denoted $|X|$. Let k be a positive integer.

The set of all subsets of the same cardinality k of the set S is denoted P_k, i.e.

$$P_k = \{\{x_{i_1}, x_{i_2}, ..., x_{i_k}\} \mid x_{i_1}, x_{i_2}, ..., x_{i_k} \in S\}.$$

Algorithm Find_the_set_of_all_coverings;
    begin
        $C := \emptyset$;
        for each attribute x in S **do**
            compute partition $\{x\}^*$;
        compute partition R^*;
        $k := 1$;
        while $k \le |S|$ **do**
            begin
                for each set P in P_k **do**
                    if (P is not a superset of any member of C) **and**

$$\left(\prod_{x \in P} \{x\}^* \le R^*\right)$$

then add P to C;
                    k := k+1
                end {while}
            end {procedure}.
            Time complexity of the algorithm for finding the set of all coverings of R in S is exponential.

3.3.6 An Algorithm for Finding a Covering

The following algorithm returns set C containing at most one covering of R in S. If the set of all coverings of R in S is empty, the algorithm returns the empty set.

Algorithm Find_one_covering;
    begin
        compute partition R*;
        compute partition S*;
        P := S;
        C := Ø;
        if S* ≤ R*
            then
                begin
                    for each attribute q in S **do**
                        begin
                            Q := P − {q};
                            compute partition Q*;
                            if Q* ≤ R*
                                then P := Q
                        end {for};
                    C := P
                end {then}
    end {procedure}.
            Time complexity of the algorithm for finding a covering of R in S is $O(m^2n)$, where m is the number of all entities, and n is the number of all attributes.

3.3.7 Rule Induction from a Decision Table

The basic tool for rule induction from a decision table is a set of coverings. The algorithms for finding coverings are given in Subsections 3.3.5 and 3.3.6. The task is to find a complete set of rules for every decision, if such a set exists.

If, for a given decision, the set of coverings is empty, then the set of rules, defined so far, does not exist. In this chapter we are not taking uncertainty into account. A method for learning rules under uncertainty is studied in Section 7.3.

Table 3.10 A Decision Table

	Attributes				Decisions			
	a	*b*	*c*	*d*	*f*	*g*	*h*	*l*
x_1	0	L	0	L	0	L	D	0
x_2	0	R	1	L	1	L	D	1
x_3	0	L	0	L	0	L	D	2
x_4	0	R	1	L	1	L	D	3
x_5	1	R	0	L	2	H	D	4
x_6	1	R	0	L	2	H	D	4
x_7	2	S	2	H	3	H	U	4
x_8	2	S	2	H	3	H	U	4

3.3.7.1 Essential Attributes

For any decision, a set of coverings may be found using the algorithm from Subsection 3.3.5. An attribute from any covering of such an decision is called *essential* for that decision. The concept of an essential attribute will be illustrated by the decision table represented by Table 3.10.

Let S be the set $\{a, b, c, d\}$ of all attributes and U be the set $\{x_1, x_2, x_3, x_4, x_5, x_6, x_7, x_8\}$ of all entities. Then the partitions of U, generated by single attributes, are:

$$\{a\}^* = \{\{x_1, x_2, x_3, x_4\}, \{x_5, x_6\}, \{x_7, x_8\}\},$$
$$\{b\}^* = \{\{x_1, x_3\}, \{x_2, x_4, x_5, x_6\}, \{x_7, x_8\}\},$$
$$\{c\}^* = \{\{x_1, x_3, x_5, x_6\}, \{x_2, x_4\}, \{x_7, x_8\}\},$$
$$\{d\}^* = \{\{x_1, x_2, x_3, x_4, x_5, x_6\}, \{x_7, x_8\}\}.$$

Let R be the set $\{f\}$. Then

$$\{f\}^* = \{\{x_1, x_3\}, \{x_2, x_4\}, \{x_5, x_6\}, \{x_7, x_8\}\},$$

and

$$\{a\}^* \not\leq \{f\}^*,$$
$$\{b\}^* \not\leq \{f\}^*,$$
$$\{c\}^* \not\leq \{f\}^*,$$
$$\{d\}^* \not\leq \{f\}^*.$$

The following may be easily checked:

$$\{a\}^*\cdot\{b\}^* \leq \{f\}^*, \quad \{a\}^*\cdot\{d\}^* \not\leq \{f\}^*,$$
$$\{a\}^*\cdot\{c\}^* \leq \{f\}^*, \quad \{b\}^*\cdot\{d\}^* \not\leq \{f\}^*,$$
$$\{b\}^*\cdot\{c\}^* \leq \{f\}^*, \quad \{c\}^*\cdot\{d\}^* \not\leq \{f\}^*.$$

The coverings of $R = \{f\}$ in $S = \{a, b, c, d\}$ are $\{a, b\}$, $\{a, c\}$, and $\{b, c\}$. Thus, attributes $a, b,$ and c are essential for f. Attribute d is not essential for f. It is not difficult to recognize that all entities, indiscernible by single attributes $a, b,$ or c, are also indiscernible by attribute d.

For $R = \{g\}$, we have

$$\{g\}^* = \{\{x_1, x_2, x_3, x_4\}, \{x_5, x_6, x_7, x_8\},$$
$$\{a\}^* \leq \{g\}^*,$$
$$\{b\}^* \not\leq \{g\}^*,$$
$$\{c\}^* \not\leq \{g\}^*,$$
$$\{d\}^* \not\leq \{g\}^*,$$
$$\{b\}^*\cdot\{c\}^* \leq \{g\}^*,$$
$$\{b\}^*\cdot\{d\}^* \not\leq \{g\}^*,$$
$$\{c\}^*\cdot\{d\}^* \not\leq \{g\}^*,$$

so coverings of $R = \{g\}$ in $S = \{a, b, c, d\}$ are $\{a\}$ and $\{b, c\}$. Here again, attributes $a, b,$ and c are essential for g.

If $R = \{h\}$, then

$$\{h\}^* = \{\{x_1, x_2, x_3, x_4, x_5, x_6\}, \{x_7, x_8\}\},$$
$$\{a\}^* \leq \{h\}^*,$$
$$\{b\}^* \leq \{h\}^*,$$
$$\{c\}^* \leq \{h\}^*,$$
$$\{d\}^* \leq \{h\}^*.$$

Coverings of $R = \{h\}$ in $S = \{a, b, c, d\}$ are $\{a\}$, $\{b\}$, $\{c\}$, and $\{d\}$. In this case, all four attributes $a, b, c,$ and d are essential for h.

For $R=\{i\}$,

$$\{i\}^* = \{\{x_1\}, \{x_2\}, \{x_3\}, \{x_4\}, \{x_5, x_6, x_7, x_8\}\}.$$

No covering of $R = \{i\}$ in $S = \{a, b, c, d\}$ exists, because

$$\{a\}^* \cdot \{b\}^* \cdot \{c\}^* \cdot \{d\}^* \not\leq \{i\}^*.$$

Thus, no attribute is essential for i.

3.3.7.2   Finding Rules from Coverings

The problem of finding rules from coverings is now discussed. The problem arises if, for a given decision, there exist two or more coverings.

As an illustration, let us consider coverings $\{a, b\}$, $\{a, c\}$, and $\{b, c\}$ for the decision f, computed previously for the decision table, presented in Table 3.10.

Any covering may be a basis for computing rules. Suppose we start from $\{a, b\}$. Then the rules are

$$(b, L) \rightarrow (f, 0),$$

$$(a, 0) \wedge (b, R) \rightarrow (f, 1),$$

$$(a, 1) \rightarrow (f, 2),$$

$$(a, 2) \rightarrow (f, 3),$$

$$(b, S) \rightarrow (f, 3).$$

The first rule, $(b, L) \rightarrow (f, 0)$, was produced by the analysis of rows x_1 and x_3 of Table 3.10. The rule $(a, 0) \wedge (b, L) \rightarrow (f, 0)$ is derived directly from the table. However, we may use the technique of dropping conditions (see 3.3.2.1) and get $(b, L) \rightarrow (f, 0)$.

A somewhat different situation arises with the last two rules, $(a, 2) \rightarrow (f, 3)$ and $(b, S) \rightarrow (f, 3)$. Both of them were obtained from rows x_7 and x_8 of Table 3.10. One may check that from the rule $(a, 2) \wedge (b, S) \rightarrow (f, 3)$ either one (but not both) conditions may be dropped, hence two rules result: $(a, 2) \rightarrow (f, 3)$ and $(b, S) \rightarrow (f, 3)$.

The preceding rules are not sufficient. Although it is possible to determine the value of the decision f, using these rules, knowing the values of attributes a and b, the set of rules can not determine f on the basis of a and c or b and c. It may happen that values of a and c are given. It is clear that the value of f can be determined because $\{f\}$ depends on $\{a, c\}$, yet there are no rules determining f from a and c. Thus, two more sets of rules are needed. The first set establishes f from a and c:

$$(a, 0) \wedge (c, 0) \rightarrow (f, 0),$$

$$(c, 1) \rightarrow (f, 1),$$

$$(a, 1) \rightarrow (f, 2),$$

$$(a, 2) \rightarrow (f, 3),$$

$$(c, 2) \rightarrow (f, 3),$$

and the second set determines f from b and c:

$$(b, L) \rightarrow (f, 0),$$
$$(c, 1) \rightarrow (f, 1),$$
$$(b, R) \wedge (c, 0) \rightarrow (f, 2),$$
$$(b, S) \rightarrow (f, 3),$$
$$(c, 2) \rightarrow (f, 3).$$

To make the analysis complete, one more observation is needed. The value 3 of the decision f may be determined from the fact that the attribute d has the value H. Thus, the rule

$$(d, H) \rightarrow (f, 3)$$

could be added to the preceding set of rules for f, in spite of the fact that d is not essential for f. Such rules will be omitted from now on, since they represent partial knowledge (d should not be used in other rules to determine the value of f, since such a rule will be subsumed by another rule). Moreover, even determining coverings is lengthy—time complexity of the algorithm for finding the set of coverings is exponential. Taking into account additional possibilities, like the preceding rule determining f from d, will increase computational complexity.

The preceding set of rules is *complete* for the decision f, i.e., the value of the decision f may be determined on the basis of all possible combinations of the essential attributes.

In the case of decision g, there are two coverings, $\{a\}$ and $\{b, c\}$. Therefore, the complete set of rules for g is

$$(a, 0) \rightarrow (g, L),$$
$$(\neg(a, 0)) \rightarrow (g, H),$$
$$(b, L) \rightarrow (g, L),$$
$$(c, 1) \rightarrow (g, L),$$
$$(b, R) \wedge (c, 0) \rightarrow (g, H),$$
$$(b, S) \rightarrow (g, H),$$
$$(c, 2) \rightarrow (g, H).$$

For the decision h, there are four coverings $\{a\}$, $\{b\}$, $\{c\}$, and $\{d\}$. Corresponding rules are skipped. This chapter is restricted to certain rules (i.e., not taking uncertainty into account); therefore, no set of rules exists for the decision i.

The goal of rule induction just presented is the selection of dependencies between coverings and decisions in order to select a *complete set of rules for all decisions*, i.e., the set of rules such that the values of all decisions, for which sets of rules are nonempty, may be determined on the basis of all possible combinations of the essential attributes.

3.3.8 Attribute Dependency and Data Bases

As was mentioned, the concept of a decision table is similar to that of a relational data base. Let us recall the basic definitions of relational data bases. A *relation scheme* is a finite set of attributes $\{a_1, a_2,..., a_n\}$. Each attribute a_i has a set of values D_i, called a *domain*. A relation r on a relation scheme is a subset of $D_1 \times D_2 \times \cdots \times D_n$. Members of r are called *tuples*.

The definition of a subset R dependent on a subset P of the set Q of all attributes and decisions (see 3.3.4.1) is analogous with the definition of functional dependency $P \to R$ of relational data bases. Even the notation is the same. The concept of a covering of R in S, for $R = S = Q$, is analogous to that of a key of a relation r on a relation scheme.

Note the differences, mentioned in Subsection 3.3.1. The specific techniques, listed previously, are not used in relational data bases. Moreover, as we will see later, decision tables are useful in some approaches of dealing with uncertainty, where they have no competition in relational data bases.

3.4 Rule-Base Verification

The first system, designed to debug the rule base, TEIRESIAS, used with MYCIN, was created by R. Davis in 1976. Other systems are ONCOCIN's rule checker (Suwa *et al.*, 1984), KES' INSPECTOR, TIMM, and CHECK, built for LES (Nguyen *et al.*, 1987).

Following Nguyen *et al.* (1987), some typical problems of rule bases, together with suggestions on how to fix them, are given. Rule-base verification includes two main groups of problems to be checked for consistency and completeness. The former group includes problems that may be checked by analyzing all pairs of rules, while the latter group consists of problems with solutions based on the analysis of the entire rule base. The assumption is that the rule syntax is restrictive enough that it is possible to compare any two rules and check for potential problems.

In the following text, only forward chaining is considered. Checking errors in backward chaining is quite similar to checking errors in forward chaining; therefore, it is omitted.

3.4.1 Consistency

Checking for consistency includes detecting redundant rules, conflicting rules, subsumed rules, rules with unnecessary conditions, and circular rules.

3.4.1.1 Redundancy

Two rules are *redundant* if and only if their condition parts are simultaneously either satisfied or not satisfied in all possible situations and their decision parts are identical.

For example, the following rules illustrate redundancy:

$$(a, L) \wedge (c, 1) \rightarrow (d, H),$$

$$(\neg(a, R)) \wedge (c, 1) \rightarrow (d, H),$$

where the attribute a may have two values: L and R. Although redundancy reduces efficiency, it should not cause serious problems in rule propagation. However, in some approaches to uncertainty, e.g., certainty factors, it may cause false results.

3.4.1.2   Conflict

Two rules are *conflicting* (or *inconsistent*) if and only if their condition parts are simultaneously satisfied or not satisfied in all possible situations and their decision parts are different for at least one situation.

The following rules are conflicting:

$$(a, L) \wedge (c, 1) \rightarrow (d, H),$$

$$(a, L) \wedge (c, 1) \rightarrow (d, L),$$

where the decision d may have one of two values: H and L.

Some rule checkers display conflicting rules and allow an expert to delete wrong rules. In some expert systems, taking uncertainty into account, there is no necessity to delete any conflicting rule.

3.4.1.3   Subsumption

One rule is *subsumed* by another if and only if one's condition part is satisfied if the other's condition part is satisfied (the converse is not true) and decision parts of both are identical.

The rule

$$(a, L) \wedge (c, 1) \rightarrow (d, H)$$

is subsumed by the rule

$$(c, 1) \rightarrow (d, H).$$

Subsumed rules cause effects similar to those of redundant rules (whenever the first rule is applicable, the second rule is also applicable).

3.4.1.4   Unnecessary Conditions

Two rules have unnecessary conditions if and only if both of them are subsumed by a third rule. The rule base may not contain this third rule, in which case checking for subsumption will not detect it.

The following two rules have unnecessary conditions:

$$(a, L) \wedge (b, D) \wedge (c, 1) \rightarrow (d, H),$$

$$(a, R) \wedge (b, D) \wedge (c, 1) \rightarrow (d, H),$$

where the attribute a may have two values: L and R. Both rules are subsumed by the rule

$$(b, D) \wedge (c, 1) \rightarrow (d, H).$$

Note that rules may have unnecessary conditions if the technique called dropping conditions (see 3.3.1.1) was either ignored or not used properly. The effects of rules with unnecessary conditions are similar to those of redundant rules.

3.4.1.5 Circularity

A set of rules is *circular* if and only if chaining of these rules results in a cycle. For example, the following three rules

$$(a, L) \wedge (c, 1) \rightarrow (d, H),$$

$$(d, H) \rightarrow (f, 2),$$

$$(f, 2) \wedge (g, 1) \rightarrow (c, 1),$$

are circular, where facts are: (a, L), $(c, 1)$, and $(g, 1)$.

With some conflict resolutions, the expert system may enter an infinite loop. One remedy to this problem is to allow the execution of any rule just once.

3.4.2 Completeness

Checking for completeness means checking for missing rules (see 3.3.1.1). Some rule checkers, e.g., the one from ONCOCIN, check if there is a rule for each possible combination of all attributes. A more recent approach (Nguyen *et al.*, 1987) checks some instances of missing rules, such as unreferenced attribute values, illegal attribute values, unreachable conditions, unreachable actions, and unreachable goals instead.

3.4.2.1 Unreferenced Attribute Values

An attribute value is *unreferenced* if and only if it is not covered by any rule's condition part. This does not necessarily mean that a rule is missing, although that is a possibility. It could be that the attribute was not essential (see 3.3.7).

3.4.2.2 Illegal Attribute and Decision Values

An attribute or decision value is *illegal* if and only if there exists a rule that refers to an attribute or decision value not in its domain. This kind of error must be corrected.

3.4.2.3 Unreachable Conditions

A rule may be used in forward chaining when its conditions are in the data base or are inferred by other rules. If neither of these requirements is fulfilled, then the rule can never be fired and its condition is *unreachable*. This error can only be identified when the class of possible initial data bases is known.

3.4.2.4   Unreachable Actions

In forward chaining, the action of a rule should match a fact from the goal or match a condition of another rule. If neither of these requirements is fulfilled, then the action is *unreachable*. In order to identify this kind of error, the class of goals must be known.

3.4.2.5   Unreachable Goals

To achieve a goal, it must be a member of the class of possible goals (which should be known) or there should exist a rule such that its action part is matched by the goal. Otherwise, the goal is *unreachable*.

3.4.3 Concluding Remarks

Despite many efforts to produce a mechanism to automatically verify knowledge in a rule base, the status quo is far from satisfactory. Although authors agree that it is necessary to develop such a mechanism, they do not agree what a subject of rule-base verification should be. Existing research is almost exclusively focused on rule-based expert systems working without uncertainty. Yet at the same time the absence of a tool for rule-base verification is recognized as an obstruction to the development of rule-based expert systems.

Exercises

1. Determine a set of rules for the following decision table:

	Attributes					Decisions		
	a	b	c	d	e	f	g	h
x_1	0	L	L	2	5	0	0	0
x_2	1	L	M	2	7	1	1	1
x_3	0	L	L	2	5	2	0	0
x_4	1	R	M	3	6	3	2	0
x_5	1	R	M	3	6	3	2	0
x_6	0	R	H	4	7	3	3	2

	Attributes					Decisions		
	a	b	c	d	e	f	g	h
x_1	0	L	L	2	5	0	0	0
x_2	1	L	M	2	7	1	1	1
x_3	0	L	L	2	5	2	0	0
x_4	1	R	M	3	6	3	2	0
x_5	1	R	M	3	6	3	2	0
x_6	0	R	H	4	7	3	3	2

2. For the following decision table:

	Attributes				Decisions			
	a	b	c	d	e	f	g	h
x_1	0	0	0	0	0	0	0	0
x_2	0	0	0	0	0	0	0	1
x_3	0	1	0	1	1	2	1	1
x_4	1	1	0	1	2	2	2	1
x_5	1	2	1	1	1	3	3	1
x_6	1	2	1	1	1	3	4	0

    a. Find decisions that can be expressed by attributes (every such decision depends on a subset of the set of attributes),

    b. For every decision that can be expressed by attributes, determine all coverings and a set of rules. Your set of rules should determine values of a decision for any possible subset of the set of attributes.

a. Find decisions that can be expressed by attributes (every such decision depends on a subset of the set of attributes),

b. For every decision that can be expressed by attributes, determine all coverings and a set of rules. Your set of rules should determine values of a decision for **any** possible subset of the set of attributes.

4. Find a decision table with three attributes, two decisions, and six entities. Both decisions should depend on different subsets of attributes. Such subsets should have at least two attributes. Find all coverings and all rules.

5. Let P and R be nonempty subsets of the set Q of all attributes and decisions. Let P^* and R^* be partitions generated by P and R, respectively. Show that

a. $P^* \cdot R^* = (P \cup R)^*$,

b. $P^* \cdot R^* \leq (P \cap R)^*$, where $P \cap R \neq \emptyset$.

6. Let P be a proper subset of the set Q of all attributes and decisions. Let P^* and Q^* be partitions generated by P and Q, respectively. Prove or give a counterexample to the following conjecture:

If for all P, $P^* \neq Q^*$, then for all $p, q \in Q$ neither $\{p\}^* \leq \{q\}^*$ nor $\{q\}^* \leq \{p\}^*$.

7. Let P be a subset of the set Q of all attributes and decisions. Let P^* and Q^* be partitions generated by P and Q, respectively. Prove or give a counterexample to the following conjecture:

If there exists a subset R of Q such that R is a proper subset of $Q - P$ and $P^* \cdot R^* = Q^*$, then there exists a proper subset S of Q such that $S^* = Q^*$.

8. Let P be a subset of the set Q of all attributes and decisions. Let P^* and Q^* be partitions generated by P and Q, respectively. Prove or give a counterexample to the following conjecture:

If there exist $p, q \in Q, p \neq q$, such that $\{p\}^* \leq \{q\}^*$ or $\{q\}^* \leq \{p\}^*$, then for every subset P of Q with $\{p, q\} \subseteq P$, there exists a proper subset R of P with $P^* = R^*$.

CHAPTER
4

ONE-VALUED QUANTITATIVE APPROACHES

This chapter begins the discussion on quantitative approaches to uncertainty in expert systems. The problem of dealing with uncertainty is crucial in the entire expert system field because in most real-life situations the expert system is forced to reason with the presence of uncertainty. Recently, the problem of reasoning under uncertainty has gained enormous attention, as is visible from the explosion of the number of monographs, journal articles, and conference papers. There have been many books published on handling uncertainty in artificial intelligence in general, and in expert systems in particular (Bouchon and Yager, 1987; Bouchon et al., 1988; Gupta and Yamakawa, 1988a, 1988b; Goodman and Nguyen, 1985; Graham and Jones, 1988; Kanal and Lemmer, 1986; Konolige, 1986; Lemmer and Kanal, 1988; Pearl, 1988; Shoham, 1988; and Smets et al., 1988). New journals concerned wholly with uncertainty in artificial intelligence have been developed.

In this chapter, three approaches for reasoning under uncertainty are introduced, based on Bayes' rule, belief networks, and certainty factors, respectively. In all three approaches, uncertainty is represented by a single number. In the first two cases, it is probability, while in the third case it is a certainty factor, which may be interpreted by probability theory as well.

4.1 Probability Theory

In this section, the most important facts from probability theory are briefly reviewed. Among all numeric approaches to uncertainty, probability theory is one of the oldest. Although probability theory was viewed recently as an inadequate model for uncertainty in artificial intelligence, many authors now think that it is the best tool.

4.1.1 Definition of Probability

The concept of probability is controversial. People perceive it differently and vigorously defend their positions. Among many approaches, one of the oldest is based on the idea of probability as a *relative frequency*, also called *objective probability*. As a result of an experiment, the relative frequency of an outcome of the experiment is the ratio of the number of occurrences of the outcome to the total number of occurrences of all possible outcomes. For example, when the experiment is throwing a die, and in 100 throws, six appeared 18 times on top, then the probability of that outcome is 0.18.

A set of possible outcomes of an experiment is called a *sample space* S. A subset of the set S is called an *event*. Thus, the power set 2^S is the *event space*, denoted E. The empty subset of S is called the impossible event. The events that contain only one element are called *simple events*. The sample space S is called a *certain event*. In the case of throwing a die, $S = \{1, 2, 3, 4, 5, 6\}$, and the event space E consists of 64 elements, ranging from the impossible event to the certain event S.

Another definition of probability is based on a *belief* of an individual about the outcome of an experiment. In this case, probabilities are *subjective* because different individuals may assign different probabilities. However, the same individual should assign the same value for the same event.

4.1.2 Kolmogorov's Axioms

A. N. Kolmogorov introduced the following system of three axioms:

Axiom 1. A real nonnegative number $P(A)$, called the probability of A, is assigned to every event A from the event space E.

Axiom 2. The probability of the certain event S is equal to 1.

Axiom 3. The addition axiom: Let $A_1, A_2,..., A_n$ be mutually exclusive events. Then

$$P(A_1 \cup A_2 \cup \cdots \cup A_n) = P(A_1) + P(A_2) + \cdots + P(A_n).$$

Note that when E is infinite, the third axiom is valid for an infinite number of mutually exclusive events.

From the preceding axioms the probability of the union of events A and B may be calculated, since

$$A \cup B = A \cup (B - (A \cap B))$$

and

$$B = (A \cap B) \cup (B - (A \cap B)).$$

Obviously, A and $B - (A \cap B)$ are disjoint, and so are $A \cap B$ and $B - (A \cap B)$. Thus, due to the addition axiom

$$P(A \cup B) = P(A) + P(B - (A \cap B))$$

and

$$P(B) = P(A \cap B) + P(B - (A \cap B)),$$

hence

$$P(A \cup B) = P(A) + P(B) - P(A \cap B).$$

Similarly, the sum of probabilities of an event A and its complement $S - A$ is equal to 1. Indeed, A and $S - A$ are disjoint, and S is the certain event, so from axioms 2 and 3

$$P(A) + P(S - A) = 1.$$

Complement $S - A$ is also denoted $\overline{A}$.

4.1.3 Conditional Probability

Suppose that the experiment is throwing a die. The sample space is $\{1, 2, 3, 4, 5, 6\}$. Suppose that additional information is given; the number X on the top of the die is greater than 2. The new sample space is $\{3, 4, 5, 6\}$. Then the probability that X is even, given that X is greater than 2 is 2/4. Say that A denotes the event that X is even and B denotes the event that X is greater than 2.

The event that X is even and greater than 2 is the intersection of A and B. Using S to compute probabilities, we conclude that the probability of A given B is $P(A \cap B)/P(B)$. The preceding motivates the following definition: Let A and B be events from the event space E. Then the *conditional probability of A given B*, denoted $P(A|B)$, is

$$\frac{P(A \cap B)}{P(B)},$$

where $P(B) > 0$. Similarly,

$$P(B|A) = \frac{P(A \cap B)}{P(A)},$$

where $P(A) > 0$. Thus

$$P(A \cap B) = P(A|B) \cdot P(B) = P(B|A) \cdot P(A).$$

Note that events may represent propositions. If propositions X, Y are represented by events A, B, then $P(\neg X)$ corresponds to $P(S - A)$, $P(X \vee Y)$ to $P(A \cup B)$, and $P(X \wedge Y)$ to $P(A \cap B)$. The joint probability $P(A \cap B)$ and its equivalent $P(X \wedge Y)$ are generally denoted as $P(A, B)$, $P(AB)$, $P(X, Y)$, and $P(XY)$.

4.1.4 Independent Events

Let A and B be events from the event space E. In general, $P(A|B)$ and $P(A)$ are different. If $P(A|B) = P(A)$, then the probability of the event A is independent of the fact that B is given, or that B happened. Events A and B are defined to be *independent* if and only if $P(A|B) = P(A)$. Moreover, $P(A|B) = P(A)$ if and only if $P(B|A) = P(B)$. Obviously, $P(A|B) = P(A)$ if and only if

$$P(A \cap B) = P(A) \cdot P(B).$$

4.1.5 Bayes' Rule

In this subsection, let $A_1, A_2,...$ be mutually disjoint events from the event space E, and let $A_1 \cup A_2 \cup \cdots = S$, where S is the sample space. Let $P(A_i) > 0$ for $i = 1, 2,....$. Let B be any event from E. Then

$$P(B) = P(A_1) \cdot P(B|A_1) + P(A_2) \cdot P(B|A_2) + \cdots.$$

The proof follows from the fact that B may be presented as the union of the following mutually disjoint subsets of the sample space S:

$$A_1 \cap B, A_2 \cap B,....$$

Thus, by the addition axiom,

$$P(B) = P(A_1 \cap B) + P(A_2 \cap B) + \cdots,$$

and, for each i,

$$P(A_i \cap B) = P(A_i) \cdot P(B|A_i).$$

Bayes' rule. If $P(B) > 0$, then

$$P(A_i|B) = \frac{P(A_i) \cdot P(B|A_i)}{P(A_1) \cdot P(B|A_1) + P(A_2) \cdot P(B|A_2) + \cdots},$$

where $i = 1, 2, ...$.

The preceding formula follows from the following fact:

$$P(A_i|B) = \frac{P(A_i \cap B)}{P(B)} = \frac{P(A_i) \cdot P(B|A_i)}{P(B)},$$

where $P(B)$ is computed earlier.

Probabilities $P(A_i)$ from the formula are called *probabilities a priori*, and $P(A_i|B)$ are *probabilities a posteriori*. Bayes' rule determines the probability of event A_i, provided event B is known.

Example. In a university that is to remain nameless, 40% of the students live in residence hall A_1, 35% live in residence hall A_2, and 25% in residence hall A_3. The event of complete satisfaction with a residence hall is denoted by B. The probability of complete satisfaction with residence halls A_1, A_2, and A_3 is $P(B|A_1) = 0.8$, $P(B|A_2) = 0.9$, and $P(B|A_3) = 0.85$, respectively. What is the probability that a completely satisfied student lives in residence hall A_1? The answer is given by Bayes' rule

$$P(A_1|B) = \frac{0.4 \cdot 0.8}{0.4 \cdot 0.8 + 0.35 \cdot 0.9 + 0.25 \cdot 0.85} = 0.378.$$

4.2 Systems Using Bayes' Rule

Bayes' rule is used in expert systems such as PROSPECTOR (Duda *et al.*, 1976) and AL/X (Reiter, 1980). In this section we follow the approach presented by R. Duda, P. Hart, and N. J. Nilsson. A useful model of propagation of knowledge in such systems is the inference network.

4.2.1 Inference Network

In production systems, the typical rule is of the form

$$E_1 \wedge E_2 \wedge \cdots \wedge E_m \rightarrow H.$$

The assumption is that all terms $E_1, E_2,..., E_m, H$ are not true or false but that they are uncertain. In the currently discussed approach, the uncertainties are represented by probabilities. For the sake of simplicity, the preceding form of the rule is simplified to the following form:

$$E \rightarrow H.$$

In order to replace the conjunction of $E_1, E_2,..., E_m$ by E, the probability of E should be computed as the joint probability of events $E_1, E_2,..., E_m$, that is,

$$P(E) = P(E_1, E_2,..., E_m).$$

An *inference network* is a directed, labeled graph in which nodes represent terms and arcs represent rules. A single rule of the form $E \rightarrow H$ is illustrated by Figure 4.1, and an example of a simple production system, consisting only of rules of the form $E \rightarrow H$, is represented by Figure 4.2. The preceding concept of inference network is a special case of the inference network from Subsection 2.2.4.

4.2.2 Bayesian Updating

The question is: What is the *a posteriori* probability of H given E for a rule $E \rightarrow H$. The answer is given by Bayes' rule:

Figure 4.1 A Rule $E \rightarrow H$

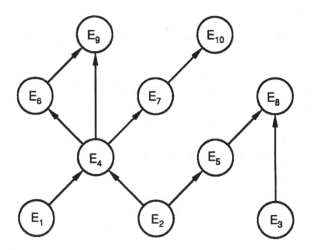

Figure 4.2 An Inference Network

$$P(H|E) = \frac{P(E|H) \cdot P(H)}{P(E)}.$$

Similarly, for a complement $\bar{H}$ of H, we have

$$P(\bar{H}|E) = \frac{P(E|\bar{H}) \cdot P(\bar{H})}{P(E)}.$$

Dividing the two equations, we obtain

$$\frac{P(H|E)}{P(\bar{H}|E)} = \frac{P(E|H)}{P(E|\bar{H})} \cdot \frac{P(H)}{P(\bar{H})}.$$

The fraction

$$\frac{P(H)}{P(\bar{H})}$$

is called the *prior odds* O(H) *on* H, the fraction

$$\frac{P(H|E)}{P(\bar{H}|E)}$$

the *posterior odds* O(H|E) *on* H *given* E, and the *likelihood ratio* λ is defined
by

$$\frac{P(E|H)}{P(E|\bar{H})}.$$

Thus, Bayes' rule has the new form

$$O(H|E) = \lambda \cdot O(H).$$

The high value of λ means that E is sufficient to determine H.
An event A having probability $P(A)$ has the following odds:

$$O(A) = \frac{P(A)}{1 - P(A)}.$$

Thus,

$$P(A) = \frac{O(A)}{1 + O(A)}.$$

Thus, the high value of $O(A)$ is transformed into high value of $P(A)$. When $O(A) = 1$ then $P(A) = 0.5$.

By analogy, we obtain

$$O(H|\bar{E}) = \bar{\lambda} \cdot O(H),$$

where

$$\bar{\lambda} = \frac{P(\bar{E}|H)}{P(\bar{E}|\bar{H})} = \frac{1 - P(E|H)}{1 - P(E|\bar{H})}.$$

The low value of $\bar{\lambda}$ means that E is necessary for determining H, because a false $\bar{E}$ reduces odds on H.

Both λ and $\bar{\lambda}$ should be given by an expert, although they are related, because

$$\bar{\lambda} = \frac{1 - P(E|H)}{1 - P(E|\bar{H})} = \frac{1 - \dfrac{P(E|H)}{P(E|\bar{H})} \cdot P(E|\bar{H})}{1 - P(E|\bar{H})} = \frac{1 - \lambda \cdot P(E|\bar{H})}{1 - P(E|\bar{H})}.$$

Similarly,

$$P(E|\bar{H}) = \frac{1 - \bar{\lambda}}{\lambda - \bar{\lambda}}$$

and

$$P(E|H) = \lambda \cdot \frac{1 - \bar{\lambda}}{\lambda - \bar{\lambda}}.$$

Thus if E is true, the odds for H given E are determined by Bayes' rule for $O(H|E)$, and if E is false, the odds for H given E are determined by Bayes' rule for $O(H|\bar{E})$. The problem is what to do if $0 < P(E) < 1$. That will be discussed in the next subsection.

Figure 4.3 Uncertain Evidence E

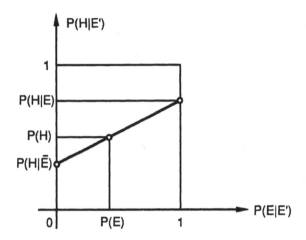

Figure 4.4 Updated Probability of H

4.2.3 Uncertain Evidence

The assumption made by R. Duda *et al.* is that if the evidence E is uncertain then the conditional probability $P(E|E')$ is given, where E' is a relevant observation (see Figure 4.3).

Thus, if E is known to be true or false, then the knowledge of E' relevant to E is no longer useful for observing H. Taking that into consideration,

$$P(H|E') = P(H, E \mid E') + P(H, \bar{E} \mid E')$$

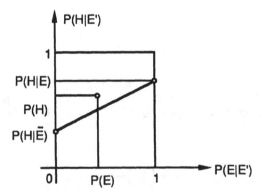

Figure 4.5 Inconsistent Prior Probabilities $P(E)$ and $P(H)$

$$= P(H \mid E, E') \cdot P(E \mid E') + P(H \mid \overline{E}, E') \cdot P(\overline{E} \mid E')$$

$$= P(H \mid E) \cdot P(E \mid E') + P(H \mid \overline{E}) \cdot P(\overline{E} \mid E')$$

$$= P(H \mid E) \cdot P(E \mid E') + (P(H \mid \overline{E}) \cdot (1 - P(E \mid E')))$$

$$= P(H \mid \overline{E}) + (P(H \mid E) - P(H \mid \overline{E})) \cdot P(E \mid E').$$

The last equation shows how the updated probability $P(H \mid E')$ depends on the current probability $P(E \mid E')$. Let us observe that if

$$P(E \mid E') = P(E)$$

then

$$P(H \mid E') = P(H, E) + P(H, \overline{E}) = P(H)$$

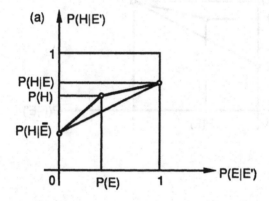

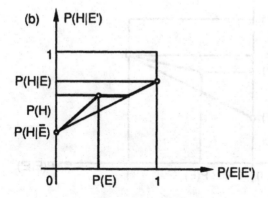

Figure 4.6 (a) and (b) Consistent Functions

(see Figure 4.4).

In practice, the expert who assigns prior probabilities in a subjective way will provide the value for $P(E)$ inconsistent with the value that follows from $P(H)$ and the linear dependency of $P(H|E')$ on $P(E|E')$ (see Figure 4.5). The values of prior probabilities should not be changed because E and H are variables, representing arbitrary nodes of the inference net (e.g., E may represent node E_4,

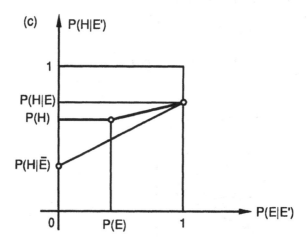

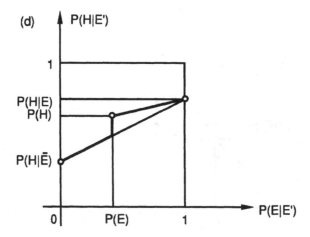

Figure 4.6 (c) and (d) Consistent Functions

while H may represent node E_7 from Figure 4.2). By making prior probabilities $P(E_4)$ and $P(E_7)$ consistent, the consistency of $P(E_4)$ with $P(E_1)$ and $P(E_2)$ will be violated.

Another possibility is to change the linear dependence of $P(H|E')$ on $P(E|E')$, so that the value for $P(E)$ is $P(H)$ for the new function. Four possibilities of adjusting the original linear function are presented in Figure 4.6.

4.2.4 Multiple Evidences and Single Hypothesis

Suppose that m evidences $E_1, E_2,..., E_m$ support the same hypothesis H using m rules $E_1 \rightarrow H, E_2 \rightarrow H,..., E_m \rightarrow H$. The additional assumption is that all pieces of evidence are conditionally independent, i.e., that

$$P(E_{j_1}, E_{j_2},..., E_{j_k} \mid H) = \prod_{i \in J} P(E_i|H)$$

and

$$P(E_{j_1},..., E_{j_2},..., E_{j_k} \mid \bar{H}) = \prod_{i \in J} P(E_i|\bar{H}),$$

where $J = \{j_1, j_2,..., j_k\}$ is an arbitrary subset of the indices 1, 2, ..., m. If all the pieces of evidence are true, then the odds on H are updated as follows:

$$O(H|E_1, E_2,..., E_m) = \left(\prod_{i=1}^{m} \lambda_i\right) \cdot O(H),$$

where

$$\lambda_i = \frac{P(E_i|H)}{P(E_i|\bar{H})} .$$

If all the pieces of evidence are false, then

$$O(H|\bar{E}_1, \bar{E}_2,..., \bar{E}_m) = \left(\prod_{i=1}^{m} \bar{\lambda}_i\right) \cdot O(H),$$

where

$$\bar{\lambda}_i = \frac{P(\bar{E}_i|H)}{P(\bar{E}_i|\bar{H})} .$$

For the general case when the evidences $E_1, E_2,..., E_m$ are neither true nor false, the effective likelihood ratio

$$\lambda_i' = \frac{O(H|E_i')}{O(H)}$$

may be computed for each $i = 1, 2,..., m$ separately, using the former subsection. With the assumption that $E_1', E_2',..., E_m'$ are mutually independent, the updating formula is

$$O(H|E_1', E_2', ..., E_m') = \left(\prod_{i=1}^{m} \lambda_i' \right) \cdot O(H).$$

4.2.5 Multiple Evidences and Multiple Hypotheses

In this subsection, it is assumed that m evidences $E_1, E_2, ..., E_m$ support n hypotheses $H_1, H_2, ..., H_n$, and that $n > 2$. The following assumptions are valid for this subsection:

1. Hypotheses $H_1, H_2, ..., H_n$ are mutually exclusive; that is,

$$P(H_i, H_j) = 0 \text{ for } i \neq j,$$

2. They are also jointly exhaustive; that is,

$$\sum_{i=1}^{n} P(H_i) = 1,$$

3. Pieces $E_1, E_2, ..., E_m$ of evidence are conditionally independent; that is,

$$P(E_{j_1}, E_{j_2}, ..., E_{j_k} | H_i) = \prod_{i \in J} P(E_i | H_i)$$

and

$$P(E_{j_1}, E_{j_2}, ..., E_{j_k} | \bar{H}_i) = \prod_{i \in J} P(E_i | \bar{H}_i),$$

where $J = \{j_1, j_2, ..., j_k\}$ is an arbitrary subset of the indices 1, 2, ..., m.

The following result was shown in (Pednault *et al.*, 1981). With the preceding assumptions, let A and B be arbitrary sets that are the intersections of some of the sets $E_1, E_2, ..., E_m$. Then

$$P(A, B) = P(A) \cdot P(B),$$

i.e., A and B are independent.

Indeed,

$$P(A, B|H_i) = P(A|H_i) \cdot P(B|H_i),$$

or

$$\frac{P(A, B, H_i)}{P(H_i)} = \frac{P(A, H_i)}{P(H_i)} \cdot \frac{P(B, H_i)}{P(H_i)}.$$

Hence,

$$(A, B, H_i) \cdot P(H_i) = P(A, H_i) \cdot P(B, H_i).$$

By analogy,

$$P(A, B, \bar{H}_i) \cdot P(\bar{H}_i) = P(A, \bar{H}_i) \cdot P(B, \bar{H}_i).$$

For any X and Y,

$$P(X, \bar{Y}) = P(X) - P(X, Y).$$

Then

$$(P(A, B) - P(A, B, H_i)) \cdot (1 - P(H_i)) = (P(A) - P(A, H_i)) \cdot (P(B) - P(B, H_i))$$

and, after multiplication,

$$P(A, B) - P(A, B, H_i) - P(A, B) \cdot P(H_i) + P(A, B, H_i) \cdot P(H_i)$$
$$= P(A) \cdot P(B) - P(A) \cdot P(B, H_i) - P(B) \cdot P(A, H_i) + P(A, H_i) \cdot$$
$$P(B, H_i),$$

or

$$P(A, B) - P(A, B, H_i) - P(A, B) \cdot P(H_i)$$
$$= P(A) \cdot P(B) - P(A) \cdot P(B, H_i) - P(B) \cdot P(A, H_i).$$

By summing over H_i, we obtain

$$n \cdot P(A, B) - P(A, B) - P(A, B) = n \cdot P(A) \cdot P(B) - P(A) \cdot P(B) - P(A) \cdot P(B),$$

or

$$(n - 2) \cdot P(A, B) = (n - 2) \cdot P(A) \cdot P(B)$$

Hence,

$$P(A, B) = P(A) \cdot P(B),$$

because $n > 2$.

The preceding result implies that A and B are independent. A and B are arbitrary intersections of some of the sets $E_1, E_2,..., E_m$; therefore,

$$P(E_{j_1}, E_{j_2},..., E_{j_k}) = P(E_{j_1}) \cdot P(E_{j_2}) \cdots P(E_{j_k}),$$

where $\{j_1, j_2,..., j_k\}$ is an arbitrary subset of the indices $1, 2, ..., m$. Thus, under assumptions (1)–(3), pieces $E_1, E_2,..., E_m$ of evidence are independent, not only conditionally independent.

Using the above result, R. W. Johnson showed in (1986) that for each hypothesis H_i there is at most one piece E_j of evidence that produces updating of the probability of H_i. This means that under assumptions (1)–(3) multiple updating is not possible at all.

4.3 Belief Networks

Belief networks, also called Bayesian networks, were introduced by J. Pearl in (1985, 1986a, 1986b), 1988). A formalism used in belief networks classifies them as a part of the Bayesian family of evidential reasoning.

Belief networks are directed acyclic graphs (i.e., without directed cycles) in which *nodes* represent variables and *arcs* represent dependencies between related variables. Such variables have values that correspond to mutually exclusive hypotheses or observations. Strengths of such dependencies are determined by conditional probabilities. An example of the belief network is presented in Figure 4.7. For a belief network representing variables, a joint probability is given by the following formula:

$$\prod_{i=1}^{n} P(x_i|S_i),$$

where S_i is the set of parents for a node x_i. In the example from Figure 4.7,

$P(x_1, x_2, x_3, x_4, x_5, x_6, x_7)$

$\qquad = P(x_1) \cdot P(x_2|x_1) \cdot P(x_3|x_1) \cdot P(x_4|x_1) \cdot P(x_5|x_2, x_3) \cdot P(x_6|x_4) \cdot$
$\qquad P(x_7|x_3, x_5).$

4.3.1 Detection of Independence

One of the useful features of belief networks is the ability to detect independence. The criterion is based on the following concepts. For a triple of nodes x, y, and z, where x is connected to z through y, two arcs are involved. Three cases are possible:

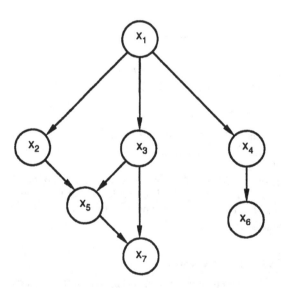

Figure 4.7 Belief Network

1. Tail-to-tail: $x \leftarrow y \rightarrow z$,

2. Head-to-tail: $x \rightarrow y \rightarrow z$ or $x \leftarrow y \leftarrow z$,

3. Head-to-head: $x \rightarrow y \leftarrow z$.

Let S be a subset of the set of all variables of a belief network. Two arcs meeting head-to-tail or tail-to-tail at node y are *blocked by* S if and only if y is in S; two arcs meeting head-to-head at node y are blocked by S if and only if neither y nor any of its descendants is in S. A path R is *separated by* S if and only if at least one pair of successive arcs along R is blocked by S. For nodes x_i and x_j, S separates x_i from x_j if and only if all paths between x_i and x_j are separated by S.

The conditional independence may be tested by the following principle: If S separates x_i from x_j then x_i is conditionally independent of x_j, given the values of variables in S. In the example from Figure 4.7, x_2 and x_3 are separated by $\{x_1\}$ but not separated by $\{x_1, x_5\}$, $\{x_1, x_7\}$, or $\{x_1, x_5, x_7\}$.

4.3.2 Knowledge Updating

In belief networks, updating is accomplished by the relaxation of constraints. An arc of the belief network represents a constraint on the possible belief value for two connected nodes. Beliefs are modified for unsatisfied constraints. In turn, the modification affects neighboring nodes so that the process continues until all constraints are satisfied.

In the case of constraint relaxation, propagation algorithms are known for

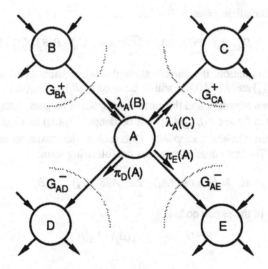

Figure 4.8 A Part of a Belief Network

some restricted classes of networks. Subsequently a propagation algorithm is described for singly connected networks. These networks are characterized by the existence of at most one path between any two nodes. Any tree is thus a singly connected network. A node A of such a network is depicted in Figure 4.8. Possible values of a node A are denoted $A_1, A_2, ..., A_n$. For any set of arcs incoming to A, a fixed conditional probability matrix is provided. For example, for arcs $B \rightarrow A$ and $C \rightarrow A$, a matrix with entries $M_{ijk} = P(A_i|B_j, C_k)$ is given.

Incoming evidence to node A through instantiated variables (i.e., data), will be denoted D. Bel(A_i) denotes the dynamic actual value of the updated node probability, that is, Bel $(A_i) = P(A_i|D)$.

The arc $B \rightarrow A$ from Figure 4.8 partitions the graph into two parts: an upper subgraph G_{BA}^+ and a lower subgraph G_{BA}^-. Data contained in G_{BA}^+ and G_{BA}^- will be denoted D_{BA}^+ and D_{BA}^- , respectively. Similarly, each of arcs $C \rightarrow A, A \rightarrow D$, and $A \rightarrow E$ partitions the graph into two subgraphs, containing corresponding data.

Thus,

$$P(D_{BA}^+, D_{CA}^+, D_{AD}^-, D_{AE}^-|A_i)$$

$$= P(D_{BA}^+, D_{CA}^+|A_i) \cdot P(D_{AC}^-|A_i) \cdot P(D_{AD}^-|A_i).$$

Using Bayes' rule,

$$\text{Bel}(A_i) = P(A_i|D_{BA}^+, D_{CA}^+, D_{AD}^-, D_{AE}^-)$$

$$= \alpha \cdot P(A_i|D_{BA}^+, D_{CA}^+) \cdot P(D_{AC}^-|A_i) \cdot P(D_{AD}^-|A_i) \cdot P(D_{AD}^-|A_i),$$

where α is a normalizing factor. Furthermore,

$$\text{Bel}(A_i) = \alpha \cdot P(D_{AC}^-|A_i) \cdot P(D_{AD}^-|A_i) \cdot (\sum_{j,k} P(A_i|B_j, C_k) \cdot P(B_j|D_{BA}^+) \cdot P(C_k|D_{CA}^+))$$

In the last equation, the current strengths of incoming arcs to A [i.e., the values $P(B_j|D_{BA}^+)$ and $P(C_k|D_{CA}^+)$] will be denoted $\pi_A(B_j)$ and $\pi_A(C_k)$, respectively, and called *casual supports*, and the current strengths of outgoing arcs from A [i.e, the values $P(D_{AC}^-|A_i)$ and $P(D_{AD}^-|A_i)$] will be denoted $\lambda_C(A_i)$ and $\lambda_D(A_i)$, respectively, and called *diagnostic supports*. Thus each arc has two associated parameters, π and λ. The last equation now has the following form:

$$\text{Bel}(A_i) = \alpha \cdot \lambda_C(A_i) \cdot \lambda_D(A_i) \cdot \sum_{j,k} P(A_i|B_j, C_k) \cdot \pi_A(B_j) \cdot \pi_A(C_k).$$

The belief of the parent node is

$$\text{Bel}(B_j) = \alpha \cdot \pi_A(B_j) \cdot \lambda_A(B_j).$$

The next question is how to update the values of parameters on the basis of values of π and λ of the neighboring arcs. The answer is given by the following two equations:

$$\lambda_A(B_i) = \alpha \cdot \sum_j (\pi_A(C_j) \cdot \sum_k \lambda_D(A_k) \cdot \lambda_E(A_k) \cdot P(A_k|B_i, C_j)),$$

and

(a) (b)

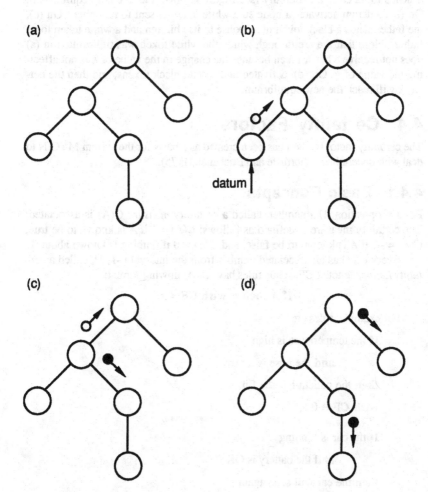

datum

(c) (d)

Figure 4.9 Belief Propagation in a Tree

$$\pi_D(A_i) = \alpha \cdot \lambda_E(A_i) \cdot \sum_{j,k} P(A_i|B_j, C_k) \cdot \pi_A(B_j) \cdot \pi_A(C_k).$$

As follows from the last two equations, a change in the value of the causal parameter π will not affect the value of the diagnostic parameter λ at the same arc, and vice versa. No circular reasoning can take place due to this fact.

Figure 4.9 depicts successive states of belief propagation through a tree. Our argumentation is similar to that of (Pearl, 1986). White tokens represent values $\lambda_A(B_j)$ that A sends to its father B; black ones represent values $\pi_D(A_i)$ that A sends to its child D. Initially [see Figure 4.9 (a)], the tree is in equilibrium. On (b), a datum activates a node so a white token is sent to its father. On (c), the father sends a black token in response to its children and a white token to its father. Note that the arc through which the white token was transmitted in (a) does not receive a black token because the change in the value of λ is not affecting the value of π. On (d), activated nodes send black tokens, and then the network will reach the new equilibrium.

4.4 Certainty Factors

The certainty factor method has been created as a basis for the system MYCIN to deal with uncertainty (Shortliffe and Buchanan, 1975).

4.4.1 Basic Concepts

For a proposition A, a number, called a *certainty measure* $C(A)$ is associated. The certainty measure is defined as follows: $C(A) = 1$ if A is known to be true, $C(A) = -1$ if A is known to be false, and $C(A) = 0$ if nothing is known about A.

Every rule has an associated number from the interval $[-1, 1]$, called a *certainty factor*, denoted CF. Thus rules have the following format:

If A **then** B **with** CF = x.

For example, rules are

If the temperature is high

      and the nose is runny,

then the patient has the flu

with CF = 0.6,

If the car is running

      and the battery is OK

then the car will start again

with CF = 0.99.

4.4.2 Propagation

For each proposition A, the initial value of certainty measure is 0. For a rule $A \rightarrow B$ with CF = x, and true A, the new certainty measure of B is equal to x. More complicated cases are described subsequently.

4.4.2.1 Rules with Certain Single Conditions

For the rule $A \rightarrow B$, the assumptions are that $C(A) = 1$ (that is, the condition A is certain) and that the present certainty factor of B is $C(B)$. Firing the rule will cause the following updating of the certainty factor of B:

$$C(B|A) = \begin{cases} C(B) + (1 - C(B)) \cdot \text{CF} & \text{if both } C(B) \text{ and CF} \\ & \text{are greater than or equal to 0,} \\ \\ C(B) + (1 + C(B)) \cdot \text{CF} & \text{if both } C(B) \text{ and CF} \\ & \text{are smaller than 0,} \\ \\ \dfrac{C(B) + \text{CF}}{1 - \min(|C(B)|, |\text{CF}|)} & \text{otherwise,} \end{cases}$$

where $|X|$ denotes the absolute value of X.

The first case may be justified as follows: If $C(B)$ is positive, then $1 - C(B)$ represents the most a rule can increase the certainty of B. That amount is multiplied by CF and added to the previous value of the certainty of B.

The preceding formula may be applied to a number of rules $A_1 \rightarrow B, A_2 \rightarrow B,..., A_m \rightarrow B$, provided $C(A_1) = C(A_2) = \cdots = C(A_m) = 1$ and all $A_1, A_2,..., A_m$ are independent, where m is a positive integer.

The formula is symmetrical with respect to $C(B)$ and CF. When it is applied to two rules, the final result does not depend on the order of rule firing.

4.4.2.2 Rules with Uncertain Single Conditions

Suppose that the rule $A \rightarrow B$, $C(A) < 1$, i.e., the condition A is uncertain. This means that A is an uncertain fact, initially in the data base, or that A is a conclusion of another rule and as such is uncertain.

The new certainty factor CF' of the rule is determined as the product of the old certainty factor CF and the certainty of the condition $C(A)$, provided $C(A)$ is positive. If $C(A)$ is not positive, the rule is not used at all. In EMYCIN a rule cannot be fired unless the certainty of its condition is greater than the threshold value equal to 0.2. That saves time by ignoring rules that are not effective (Buchanan and Duda, 1983).

4.4.2.3 Multiple Conditions

Rules may have some conditions; for example,

If A and B and C then D with CF = 0.7,

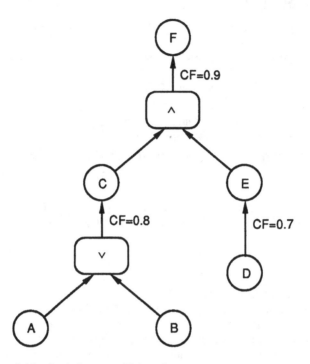

Figure 4.10 An Inference Network

If (A or not B) **and** C **and** not D **then** E **with** CF = 0.4,

If A or B or C **then** D **with** CF = 0.95.

The designers of MYCIN used the fuzzy approach to determine the certainty value of the condition part of the rule with multiple conditions. Thus, the certainty value of the condition part may be computed by repeatedly using the following formulas:

$$C(A \wedge B) = \min\ (C(A), C(B)),$$
$$C(A \vee B) = \max\ (C(A), C(B)),$$
$$C(\neg A) = 1 - C(A).$$

An example of the inference net is given in Figure 4.10. Suppose that the corresponding rules are as follows:

R1. **If** A or B **then** C **with** CF = 0.8,

R2. **If** D **then** E **with** CF = 0.7,

R3. **If C and E then F with CF = 0.9,**

and that the initial values for propositions are $C(A) = 0.4$, $C(B) = 0.6$, $C(D) = 0.9$, $C(F) = 0.2$, and $C(C) = C(E) = 0$. Then the certainty of $A \vee B$ is

$$C(A \vee B) = \max (C(A), C(B))$$
$$= \max (0.4, 0.6)$$
$$= 0.6.$$

The new certainty factor CF' for rule R1 is determined as follows:

$$CF' = CF \cdot C(A \vee B)$$
$$= 0.8 \cdot 0.6$$
$$= 0.48.$$

The new certainty factor CF' for rule R2 is

$$CF' = CF \cdot C(D) = 0.9 \cdot 0.7 = 0.63$$

The certainty of C is

$$C(C|A \vee B) = C(C) + (1 - C(C)) \cdot CF'$$
$$= 0 + (1 - 0) \cdot 0.48$$
$$= 0.48.$$

The certainty of E is

$$C(E|D) = C(E) + (1 - C(E)) \cdot CF'$$
$$= 0 + (1 - 0) \cdot 0.63$$
$$= 0.63.$$

The certainty of $C \wedge E$ is

$$C(C \wedge E) = \min (C(C), C(E))$$
$$= \min (0.48, 0.63)$$
$$= 0.48.$$

The new certainty factor CF' for rule R3 is

$$CF' = CF \cdot C(C \wedge E) = 0.9 \cdot 0.48 = 0.432.$$

Finally, the certainty of F is

$$C(F|C \wedge E) = C(F) + (1 - C(F)) \cdot CF'$$
$$= 0.2 + (1 - 0.2) \cdot 0.432$$
$$= 0.546.$$

4.5 Concluding Remarks

According to extreme opinions (Cheeseman, 1985), a subjective method of probability theory is all that is needed, excluding any other theory for reasoning under uncertainty, such as the Dempster-Shafer theory, fuzzy logic, and so on. Obviously, many people do not share this opinion. For example, L. A. Zadeh claims that probability theory is insufficiently expressive to represent all kinds of uncertainties (Zadeh, 1986b). The subjective (or Bayesian) approach is based on the definition of probability as a measure of belief rather than relative frequency. The latter is rejected (Cheeseman, 1985) since the area of applications of such an approach is limited because of the necessity of large samples. However, the Bayesian approach is not unique in being based on belief—the same assumption is fundamental in the Dempster-Shafer theory. Some of the proponents of probability theory do not condemn objective probabilities (that is, probabilities based on frequencies known to hold in a sequence of repeatable events); see (Kyburg, 1987a).

A priori probabilities are given mostly as subjective estimates. The assumption about conditional independence, or simply independence, necessary from the point of view of computational complexity, is hard to justify. Once accepted, it leads to wrong results (see the critique of that assumption in Szolovits and Parker, 1978).

Any change in the number of pieces of evidence or hypotheses causes a lot of additional computations of values for prior odds and likelihood ratios. Thus, it is difficult to modify systems based on the Bayesian approach. A single value, representing evidence for and against the hypothesis, is not a good representative of uncertainty either. Many authors consider the fact that the single value for probability does not have specified precision a serious disadvantage.

It should be noted that there exists a method for updating a probability distribution on the basis of the *maximum entropy* (called also *least information*), applied to expert systems for the first time by K. Konolige in 1979 and then further developed in Cheeseman (1983), Hunter (1986), Shore (1986) and other works.

For some applications with well-documented and justified needs for the Bayesian approach, the approach works well, as the few hundred years' history of the subject shows. However, even in existing systems, as the preceding discussion of PROSPECTOR shows, there are problems with the tool. PROSPECTOR is only loosely based on the Bayesian approach. According to Johnson's observation, cited in Subsection 4.2.5, in systems with at least three hypotheses satisfying PROSPECTOR assumptions, at most one of the pieces of evidence can alter the probability of a hypothesis. Thus, updating is possible but is extremely restricted because multiple updating of any of the hypotheses is excluded.

J. Pearl in his book (Pearl, 1988) quotes from Perez and Jirousek (1985) a taxonomy of intelligent systems reasoning under uncertainty. According to this taxonomy, systems may be classified as *extensional* or *intensional*. In the first class of systems, uncertainty is represented and propagated in a modular way. Classical examples are MYCIN and PROSPECTOR. MYCIN is an extensional system because every time its rules are executed, local computations are performed and new values of certainties and certainty factors, for *involved* facts and rules, respectively, are computed. The second class of systems is characterized by global values and computations. Examples are systems based on probability theory (for example, INES). Probabilities are assigned to all simple events from the sample space S and are restricted by global laws; for example,

$$\sum_{x \in S} P(x) = 1.$$

Thus, if a probability of a simple event should be updated, some of the remaining probabilities must be changed as well. Belief networks are also intensional.

Lack of sound supporting theory for certainty factors is considered a disadvantage. Certainty factors, with some modifications, may be interpreted by probability theory (Heckerman, 1986). However, pieces of evidence must be conditionally independent given the hypothesis and its negation, and the inference network must have a tree structure. All these assumptions are seldom satisfied in the real world.

Exercises

1. A player casts some dice. On none of the dice does "six" appear on top. What is the smallest number of dice cast so that the probability of this event does not exceed $\frac{1}{4}$.

2. A student should learn 16 problems. He has managed to master only 10 problems. What is the probability that he will be able to solve three problems during a test?

3. Show that

$$P(A, B, C) = P(A) \cdot P(B|A) \cdot P(C|A, B).$$

4. Three students were solving problems. Student A solved 10 problems; 8 solutions were correct. Student B solved 8 problems; 6 solutions were correct. Student C solved 6 problems; 3 solutions were correct. A randomly selected solution from the set of all solutions is correct. What is the probability that it was solved by student C?

5. Hypotheses H_1, H_2, and H_3 and events E_1 and E_2 are described by the following probabilities:

$$P(H_i) = \frac{1}{3} \text{ for } i = 1, 2, 3,$$

$P(H_i, H_j) = 0$ for $i \neq j$,

$P(E_1| H_i) = \frac{1}{2}$ for $i = 1, 2, 3$,

$P(E_2| H_1) = \frac{1}{2}$,

$P(E_2| H_2) = \frac{1}{3}$,

$P(E_2| H_3) = \frac{1}{6}$,

$P(E_1, E_2|H_1) = \frac{1}{4}$,

$P(E_1, E_2|H_2) = \frac{1}{6}$,

$P(E_1, E_2|H_3) = \frac{1}{12}$.

a. Verify conditional independence under each hypothesis and its complement for both events E_1 and E_2,

b. Check whether updating occurs.

6. In the Bayesian approach to combine conditionally independent evidence, the following scheme is used:

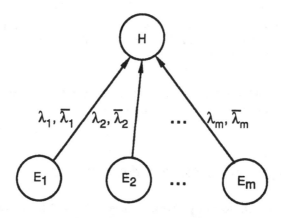

where λ_i and $\overline{\lambda}_i$ are defined in 4.2.4, and $i = 1, 2,..., m$. In the problem the assumption is that $m = 3$ and that all E_1, E_2, and E_3 are true or false, so that no interpolation is used.

a. Using the conditional independence assumption, express $O(H|E_1, E_2)$ in terms of λ's, $\overline{\lambda}$'s, and $O(H)$.

b. Dropping the assumption about conditional independence, tell what the new scheme to update the value of $O(H)$ is if single evidences and their complements, pairs of evidences and their complements, and triples of evidences and their complements are given. Draw a scheme similar to the preceding one, labeling new links by appropriate new λ's and $\bar{\lambda}$'s.

7. For the following inference net,

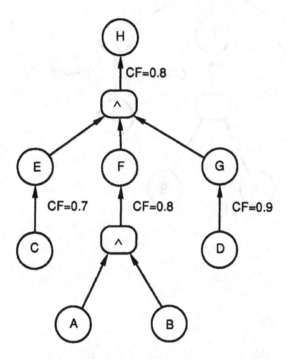

and $C(A) = 0.8$, $C(B) = 0.9$, $C(C) = 0.5$, $C(D) = 0.4$, $C(E) = C(F) = C(G) = 0$, $C(H) = 0.5$, tell the certainty of H.

8. For the following inference net, where $C(A) = 0.9$, $C(B) = C(C) = C(D) = 0.8$, $C(E) = 0.6$, $C(F) = 0.2$, and $C(G) = 0$, tell what the minimum value for the unknown certainty factor CF of the rule *If A and B then E with* CF must be if the updated value for $C(G|D \lor E \lor F)$ should be equal to 0.7.

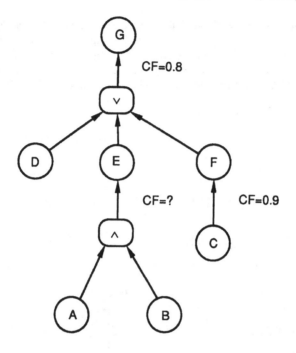

CHAPTER
5

TWO-VALUED QUANTITATIVE APPROACHES

In this chapter, two approaches for reasoning under uncertainty, the Dempster-Shafer theory and the INFERNO system, are discussed. Both are characterized by two numbers, lower and upper bounds, describing uncertainty. The first ideas of the Dempster-Shafer theory, called also the belief function theory or evidence theory, appeared in the seventeenth century, in works of G. Hooper and J. Bernoulli. The INERNO system was introduced in Quinlan (1983b). A number of systems, reasoning under uncertainty using the Dempster-Shafer theory, have been implemented, for example, Lowrance *et al.* (1986), Biswas and Anand (1987), Zarley *et al.* (1988).

5.1   Dempster-Shafer Theory

The Dempster-Shafer theory is based on the idea of placing a number between zero and one to indicate the degree of belief of evidence for a proposition (Shafer, 1976). The theory also includes reasoning based on the rule of combination of degrees of belief based on different evidences.

The addition axiom, one of the axioms of Bayesian theory, gives as a result that $P(A) + P(\overline{A}) = 1$ for any proposition A. That does not necessarily correspond to the description of the real world because ignorance was not taken into account. If we have no evidence at all, for or against A, then it is appropriate to assume that both degrees of belief, for an evidence of A and $\overline{A}$, are equal to zero.

5.1.1 Frame of Discernment

Suppose Θ is a finite nonempty set, called a *frame of discernment*.

For a subset A of Θ, a unique proposition P over Θ is defined in the following way: For an element x of Θ the value of $P(x)$ is true if and only if x is in A. Such a proposition P is *discerned* by frame Θ. Let A and B be subsets of Θ and let P and R be the corresponding propositions. Then A is the set-theoretic complement of B if and only if P is the negation of R, $A \cup B$ corresponds to $P \vee R$, and $A \cap B$ corresponds to $P \wedge R$.

5.1.2 Basic Probability Numbers

One of the basic concepts of the Dempster-Shafer theory is that of a *basic probability assignment*, that is, a function $m\colon 2^\Theta \to [0, 1]$ such that

(1) $m(\emptyset) = 0$

and

(2) $\displaystyle\sum_{A \subseteq \Theta} m(A) = 1.$

The number $m(A)$ is called a *basic probability number of* A. Condition (1) states that no belief is committed to the empty set, and (2) states that the total belief is equal to one.

5.1.3 Belief Functions

Let m be a given basic probability assignment. A function Bel: $2^\Theta \to [0, 1]$ is called a *belief function over* Θ if and only if

$$\text{Bel}\,(A) = \sum_{B \subseteq A} m(B).$$

A belief function may be characterized independently of the concept of the basic probability assignment in the following way.

Let Θ be a frame of discernment. A function Bel: $2^\Theta \to [0, 1]$ is a belief function if and only if

1. Bel $(\emptyset) = 0$,

2. Bel $(\Theta) = 1$,

3. For every positive integer n and all subsets $A_1, A_2,..., A_n$ of Θ,

$$\text{Bel}\left(\bigcup_{i \in \{1,2,...,n\}} A_i\right) \geq \sum_{\emptyset \neq I \subseteq \{1,2,...,n\}} (-1)^{|I|+1} \cdot \text{Bel}\left(\bigcap_{i \in I} A_i\right).$$

Given a belief function, a basic probability assignment is unique and may be computed from the following formula:

$$m(A) = \sum_{B \subseteq A} (-1)^{|A-B|} \cdot \text{Bel}(B),$$

where $A \subseteq \Theta$.

If a proposition A implies proposition B, then

$$\text{Bel}(A) \leq \text{Bel}(B).$$

5.1.4 Focal Elements and Core

A subset A of the frame of discernment Θ is called a *focal element of a belief function over Θ* if and only if

$$m(A) > 0,$$

where m is the basic probability assignment associated with Bel.

The union of all the focal elements of a belief function is said to be its *core*.

5.1.5 Degrees of Doubt and Plausibility

Let Bel be a belief function over Θ and let $A \subseteq \Theta$. Then the *doubt function*, denoted Dou, is a function of $2^\Theta \rightarrow [0, 1]$, defined as follows:

$$\text{Dou}(A) = \text{Bel}(\overline{A}).$$

The *plausibility function*, denoted Pl, is a function of $2^\Theta \rightarrow [0, 1]$ defined by

$$\text{Pl}(A) = 1 - \text{Dou}(A).$$

The plausibility function Pl (A) may be expressed in terms of the basic probability assignment m of Bel in the following way:

$$\text{Pl}(A) = 1 - \text{Dou}(A) = 1 - \text{Bel}(\overline{A}) = \sum_{B \subseteq \Theta} m(B) - \sum_{B \subseteq \overline{A}} m(B) = \sum_{B \cap A \neq \emptyset} m(B).$$

Belief and plausibility functions have the following properties:

$$\text{Bel}(\emptyset) = \text{Pl}(\emptyset) = 0,$$

$$\text{Bel}(\Theta) = \text{Pl}(\Theta) = 1,$$

$$\text{Bel}(A) \leq \text{Pl}(A),$$

$$\text{Bel}(A) + \text{Bel}(\overline{A}) \leq 1,$$

$$\text{Pl}(A) + \text{Pl}(\overline{A}) \geq 1,$$

if $A \subseteq B$ then $\text{Bel}(A) \leq \text{Bel}(B)$ and $\text{Pl}(A) \leq \text{Pl}(B)$.

Suppose Θ is the following set: $\{x_1, x_2, x_3, x_4, x_5\}$. Suppose the basic probability assignment m is given by:

$$m(\{x_1, x_2, x_3\}) = \frac{1}{2},$$

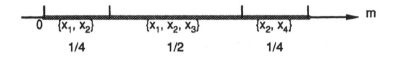

Figure 5.1 Basic Probability Assignment m

$$m\left(\{x_1, x_2\}\right) = \tfrac{1}{4},$$

$$m\left(\{x_2, x_4\}\right) = \tfrac{1}{4},$$

and

$$m(A) = 0$$

for all other $A \subseteq \Theta$. The function m may be depicted as line subsegments of the interval [0, 1] (see Figure 5.1).

Then the belief function Bel over Θ, associated with m, is defined by

$$\mathrm{Bel}\left(\{x_1, x_2\}\right) = \tfrac{1}{4},$$

$$\mathrm{Bel}\left(\{x_1, x_2, x_3\}\right) = \tfrac{3}{4},$$

$$\mathrm{Bel}\left(\{x_1, x_2, x_3, x_4\}\right) = 1,$$

$$\mathrm{Bel}\left(\{x_1, x_2, x_3, x_4, x_5\}\right) = 1,$$

$$\mathrm{Bel}\left(\{x_1, x_2, x_3, x_5\}\right) = \tfrac{3}{4},$$

$$\mathrm{Bel}\left(\{x_1, x_2, x_4\}\right) = \tfrac{1}{2},$$

$$\mathrm{Bel}\left(\{x_1, x_2, x_4, x_5\}\right) = \tfrac{1}{2},$$

$$\mathrm{Bel}\left(\{x_1, x_2, x_5\}\right) = \tfrac{1}{4},$$

$$\mathrm{Bel}\left(\{x_2, x_3, x_4\}\right) = \tfrac{1}{4},$$

$$\mathrm{Bel}\left(\{x_2, x_3, x_4, x_5\}\right) = \tfrac{1}{4},$$

$$\mathrm{Bel}\left(\{x_2, x_4\}\right) = \tfrac{1}{4},$$

$$\mathrm{Bel}\left(\{x_2, x_4, x_5\}\right) = \tfrac{1}{4},$$

and

$$\mathrm{Bel}\,(A) = 0$$

for all other $A \subseteq \Theta$.

The plausibility function Pl over Θ , associated with m, is defined by

$$Pl\ (\emptyset) = 0,$$

$$Pl\ (\{x_1\}) = \frac{3}{4},$$

$$Pl\ (\{x_1, x_3\}) = \frac{3}{4},$$

$$Pl\ (\{x_1, x_3, x_5\}) = \frac{3}{4},$$

$$Pl\ (\{x_1, x_5\}) = \frac{3}{4},$$

$$Pl\ (\{x_3\}) = \frac{1}{2},$$

$$Pl\ (\{x_3, x_4\}) = \frac{3}{4},$$

$$Pl\ (\{x_3, x_4, x_5\}) = \frac{3}{4},$$

$$Pl\ (\{x_3, x_5\}) = \frac{1}{2},$$

$$Pl\ (\{x_4\}) = \frac{1}{4},$$

$$Pl\ (\{x_4, x_5\}) = \frac{1}{4},$$

$$Pl\ (\{x_5\}) = 0,$$

and

$$Pl\ (A) = 1$$

for all other $A \subseteq \Theta$.

Focal elements of Bel are $\{x_1, x_2\}$, $\{x_1, x_2, x_3\}$, and $\{x_2, x_4\}$. The core of Bel is $\{x_1, x_2, x_3, x_4\}$.

5.1.6 Bayesian Belief Functions

Let Θ be a frame of discernment. A function Bel: $2^\Theta \to [0, 1]$ is called a *Bayesian belief function* if and only if

1. Bel $(\emptyset) = 0$,
2. Bel $(\Theta) = 1$,
3. Bel $(A \cup B) = $ Bel $(A) + $ Bel (B),

where $A, B \subseteq \Theta$ and $A \cap B = \emptyset$.

Any Bayesian belief function is a belief function. The following conditions are all equivalent

1. Bel is Bayesian,
2. All focal elements of Bel are singletons,

3. Bel = Pl,

4. Bel (A) + Bel $(\overline{A})$ = 1 for all $A \subseteq \Theta$.

Thus, if Bel is a Bayesian belief function, its basic probability assignment P is defined as a function $\Theta \rightarrow [0, 1]$ such that

1. $\displaystyle\sum_{\theta \in \Theta} P(\theta) = 1$

and

2. Bel $(A) = \displaystyle\sum_{\theta \in A} P(\theta)$

for all $A \subseteq \Theta$.

Therefore, a Bayesian belief function satisfies all three of Kolmogorov's axioms (see 5.1.2), while a belief function satisfies axioms 1 and 2, but not necessarily the addition axiom. The Bayesian belief functions are a subclass of the class of belief functions.

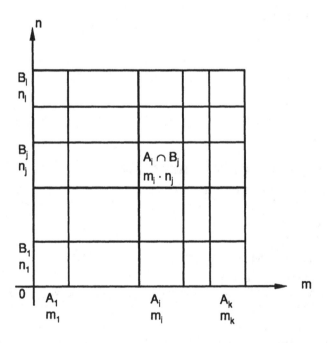

Figure 5.2 Orthogonal Sum of m and n

5.1.7 Dempster's Rule of Combination

Suppose that two different belief functions Bel_1 and Bel_2 over the same frame of discernment Θ represent different pieces of evidence. **The assumption is that both pieces of evidence are independent.** As a result of Dempster's rule of combination, a new belief function, based on the combined device, is computed as their orthogonal sum, denoted $Bel_1 \oplus Bel_2$. Bayes' rule (see 4.1.5) is a special case of Dempster's rule of combination, as shown by G. Shafer (1976).

Dempster's rule of combination is presented for basic probability assignments (there is a one to one correspondence between basic probability assignments and belief functions). Suppose that two basic probability assignments m and n are given. They are orthogonally combined and presented by Figure 5.2.

Focal elements of m are $A_1,..., A_i,..., A_k$ and focal elements of n are $B_1,..., B_j,..., B_l$. Furthermore,

$$m(A_i) = m_i$$

for $i = 1, 2,..., k$ and

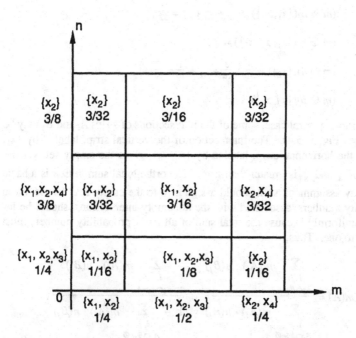

Figure 5.3 Orthogonal Sum of m and n; the Intersection of Any Two Strips Is Nonempty

$$n\left(B_j\right) = n_j$$

for $j = 1, 2,..., l$. In Figure 5.2, a vertical strip of height equal to 1, labeled by A_i, represents the basic probability number, assigned to A_i. Similarly, a horizontal strip of length equal to 1, labeled by B_j, represents the basic probability number n_j, assigned to B_j.

Provided that the two pieces of evidence are independent, the intersection of those two strips, a rectangle, labeled by $A_i \cap B_j$, has the basic probability assignment equal to $m_i \cdot n_j$.

A subset A of Θ may be represented by a few different rectangles. Then the total basic probability number committed to A by the orthogonal sum of m and n is

$$\sum_{\substack{i,j \\ A_i \cap B_j}} m(A_i) \cdot n(B_j).$$

The concept of Dempster's rule of combination, for a special case in which the intersection of any two strips, A_i and B_j, is nonempty, is illustrated by an example from Figure 5.3, in which

$$(m \oplus n)(\{x_1, x_2\}) = \tfrac{3}{32} + \tfrac{3}{16} + \tfrac{1}{16} = \tfrac{11}{32},$$

$$(m \oplus n)(\{x_1, x_2, x_3\}) = \tfrac{1}{8},$$

$$(m \oplus n)(\{x_2\}) = \tfrac{3}{32} + \tfrac{3}{16} + \tfrac{3}{32} + \tfrac{1}{16} = \tfrac{7}{16},$$

$$(m \oplus n)(\{x_2, x_4\}) = \tfrac{3}{32}.$$

In a more general case, some of the intersections of strips A_i and B_j may be empty (see Figure 5.4). The intersection of the vertical strip, labeled by $\{x_1, x_2\}$, and the horizontal strip, labeled by $\{x_4, x_5\}$, gives the empty set, yet the product of $\tfrac{1}{4}$ and $\tfrac{3}{8}$ is greater than zero. The orthogonal sum $m \oplus n$ is a basic probability assignment, so it should assign zero to the empty set. Hence, basic probability numbers, associated with the nonempty intersections, should be increased uniformly because the total sum of all basic probability numbers must be equal to one. Thus,

$$m(A) = \frac{\displaystyle\sum_{\substack{i,j \\ A_i \cap B_j = A}} m(A_i) \cdot n(B_j)}{\displaystyle\sum_{\substack{i,j \\ A_i \cap B_j \neq \emptyset}} m(A_i) \cdot n(B_j)} = \frac{\displaystyle\sum_{\substack{i,j \\ A_i \cap B_j = A}} m(A_i) \cdot \text{n}(B_j)}{1 - \displaystyle\sum_{\substack{i,j \\ A_i \cap B_j = \emptyset}} m(A_i) \cdot n(B_j)}.$$

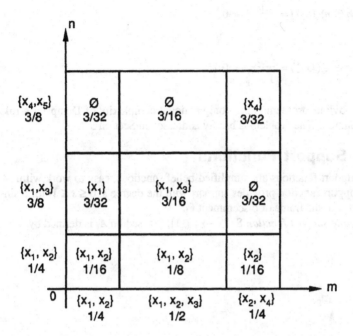

Figure 5.4 Orthogonal Sum of *m* and *n*; Some of Intersections of Vertical and Horizontal Strips Are Empty

In the example from Figure 5.4,

$$\sum_{\substack{i,j \\ A_i \cap B_j = \emptyset}} m(A_i) \cdot n(B_j) = \frac{3}{32} + \frac{3}{16} + \frac{3}{32} = \frac{3}{8},$$

$$(m \oplus n)(\{x_1\}) = \frac{\frac{3}{32}}{1 - \frac{3}{8}} = 0.15,$$

$$(m \oplus n)(\{x_1, x_2\}) = \frac{\frac{1}{16} + \frac{1}{8}}{1 - \frac{3}{8}} = 0.3,$$

$$(m \oplus n)(\{x_1, x_3\}) = \frac{\frac{3}{16}}{1 - \frac{3}{8}} = 0.3,$$

$$(m \oplus n)(\{x_2\}) = \frac{\frac{1}{16}}{1 - \frac{3}{8}} = 0.1,$$

$$(m \oplus n)(\{x_4\}) = \frac{\frac{3}{32}}{1 - \frac{3}{8}} = 0.15.$$

The obvious problem is the computational complexity of Dempster's rule of combination. The problem is briefly addressed in Section 5.3.

5.1.8 Support Functions

Simple support functions are simplified belief functions, easy to work with. A simple support function provides support with the degree s, $0 \le s \le 1$, for a single subset A of the frame of discernment Θ.

A *simple support function* S: $2^\Theta \to [0, 1]$, focused on A, is defined by

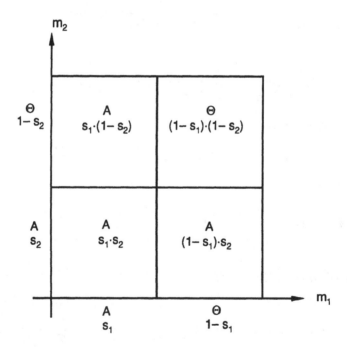

Figure 5.5 Homogenous Evidence

$$S(B) = \begin{cases} 0 \text{ if } B \text{ does not contain } A, \\[6pt] s \text{ if } B \text{ contains } A \text{ but } B \neq \Theta, \\[6pt] 1 \text{ if } B = \Theta, \end{cases}$$

where s, a *degree of support* (or *belief*) *for* A, is a number such that $0 \leq s \leq 1$.

If S is a simple support function, then S is the belief function with two focal elements, A and Θ, and with the basic probability assignment m defined by $m(A) = s$, $m(\Theta) = 1 - s$, and $m(B) = 0$ for all other $B \subseteq \Theta$.

A belief function is a *separable support function* if and only if it is a simple support function or it is equal to the orthogonal sum of two or more simple support functions. If S is a separable support function, A and B are focal elements of S, and $A \cap B \neq \emptyset$, then $A \cap B$ is a focal element of S.

The separable functions are included in a bigger class of belief functions, called *support functions*. The support functions are obtained from separable

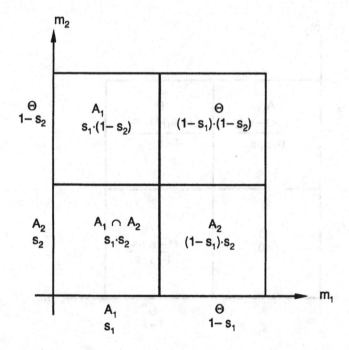

Figure 5.6 Heterogenous Evidence

support functions by coarsening the frame of discernment. A belief function is a support function if and only if its core has a positive basic probability number.

5.1.9 Combining Simple Support Functions

Suppose two simple support functions S_1 and S_2, focused on A_1 and A_2, are given, with degrees of support s_1 and s_2, respectively. The assumption is that S_1 and S_2 represent independent pieces of evidence. In combining S_1 and S_2, three cases are distinguished:

1. Homogenous evidence, when $A_1 = A_2$,

2. Heterogenous evidence, $A_1 \neq A_2$ but $A_1 \cap A_2 \neq \emptyset$,

3. Conflicting evidence, $A_1 \cap A_2 = \emptyset$.

The homogenous case is represented by Figure 5.5. It is clear that

$$(m_1 \oplus m_2)(A) = s_1 \cdot s_2 + (1 - s_1) \cdot s_2 + s_1 \cdot (1 - s_2) = s_1 + s_2 - s_1 \cdot s_2.$$

The basic probability assignment $m_1 \oplus m_2$ has the same focal elements, A and Θ, as m_1 or m_2.

In the heterogenous case, $m_1 \oplus m_2$ has four focal elements $A_1 \cap A_2, A_1,$

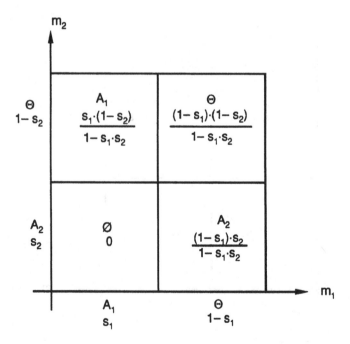

Figure 5.7 Conflicting Evidence

A_2, and Θ, if neither $A_1 \subseteq A_2$ nor $A_2 \subseteq A_1$, or three focal elements otherwise. Their basic probability numbers are represented by Figure 5.6.

The case of conflicting evidence is depicted by Figure 5.7. The basic probability assignment has three focal elements A_1, A_2, and Θ.

5.2 INFERNO

This approach to dealing with uncertainty in expert systems has been developed by J. Ross Quinlan of the Rand Corporation. INFERNO represents a conservative and cautious attitude. It has guaranteed the validity of all inferences, and—what is more controversial—has forced consistency of the information, instead of taking it into account. Among the features of this system are the ability to deal with dependent pieces of evidence and the propagation of the information through the entire system, not only toward a goal.

5.2.1 Representation of Uncertainty

A piece of evidence or a hypothesis is called a proposition. Proposition A is characterized by two values, $t(A)$ and $f(A)$, where $t(A)$ is a lower bound on the probability $P(A)$ of A, derived from the evidence for A, and $f(A)$ is a lower bound on $P(\overline{A})$, derived from the evidence against A, that is,

$$P(A) \geq t(A)$$

and

$$P(\overline{A}) \geq f(A).$$

The information about proposition A is said to be *consistent* if and only if

$$t(A) + f(A) \leq 1.$$

5.2.2 Relations

INFERNO uses its own terminology. In particular, the following *relations* between propositions are defined:

A **enables** S **with strength** X, where X is interpreted as the lower bound for the conditional probability $P(S|A)$ [i.e., $P(S|A) \geq X$],

A **inhibits** S **with strength** X, with the interpretation $P(\overline{S}|A) \geq X$,

A **requires** S **with strength** X [i.e., $P(\overline{A}|\overline{S}) \geq X$],

A **unless** S **with strength** X [i.e., $P(A|\overline{S}) \geq X$],

A **negates** S, or $A = \overline{S}$,

A **disjoins** $\{S_1, S_2,..., S_n\}$, or $A = \bigcup_i S_i$,

A **disjoins-independent** $\{S_1, S_2,..., S_n\}$, or $A = \bigcup_i S_i$ and $P(S_i \cap S_j) = P(S_i) \cdot P(S_j)$ for all $i \neq j$,

A **disjoins-exclusive** $\{S_1, S_2,..., S_n\}$, or $A = \bigcup_i S_i$ and $P(S_i \cap S_j) = 0$ for all $i \neq j$,

A **conjoins** $\{S_1, S_2,..., S_n\}$, or $A = \bigcap_i S_i$,

A **conjoins-independent** $\{S_1, S_2,..., S_n\}$, or $A = \bigcap_i S_i$ and $P(S_i \cap S_j) = P(S_i) \cdot P(S_j)$ for all $i \neq j$,

$\{S_1, S_2,..., S_n\}$ **mutually exclusive**, meaning that $P(S_i \cap S_j) = 0$ for all $i \neq j$.

5.2.3 Propagation

For each proposition A, initial values of bounds $t(A)$ and $f(A)$ are both equal to 0. Values of bounds may be changed by explicit information given to the system or by an inference.

The system is furnished with an inference net, like PROSPECTOR. The difference is that INFERNO's inference net consists of more operations on propositions (these operations are listed as relations in the previous subsection).

Simple examples of such inference nets are given by Figures 5.8 and 5.9. Each relation is provided with *constraints*, the left sides of which are bounds, while the right sides are expressions, constructed from other bounds.

Suppose that the value for a bound of proposition A was changed. All constraints associated with relations in which proposition A is involved must be checked. If the value of the right side of a constraint, computed on the basis of the new information about A, is bigger than the value of the left side of the constraint, then the bound from the left side must be increased to this new value. Such a constraint is said to be *activated*.

Any increase in the value of a bound for a new proposition must also be propagated.

5.2.4 Constraints

INFERNO propagation constraints are derived from the following basic properties:

$$P(Z) + P(\bar{Z}) = 1$$

$$t(Z) \leq P(Z) \leq 1 - f(Z)$$

for any proposition Z, and

$$\max P(S_i) \leq P(S_1 \cup S_2 \cup \cdots \cup S_n) \leq \sum_i P(S_i)$$

for all propositions $S_1, S_2,..., S_n$. The constraints for the relation A **enables** S **with strength** X are derived as follows. From the definition of the relation,

$$P(S|A) \geq X,$$

or

$$P(A) \cdot P(S|A) \geq P(A) \cdot X.$$

But

$$P(A) \cdot P(S|A) = P(S \cap A),$$

and

$$P(S) \geq P(S \cap A),$$

hence,

$$P(S) \geq P(A) \cdot X \geq t(A) \cdot X.$$

On the other hand,

$$P(S) \geq t(S).$$

Thus, the last two inequalities describe the bounds for the same probability $P(S)$ of proposition S. Value $t(S)$ is given, $t(A)$ is subject to change. If the new value of $t(A) \cdot X$ is bigger than $t(S)$, then it better describes the lower bound for $P(A)$, so that the value of $t(S)$ should be increased to $t(A) \cdot X$. Therefore, the constraint is

$$t(S) \geq t(A) \cdot X.$$

The other constraint may be derived from

$$P(A) \leq \frac{P(S)}{X},$$

or

$$\frac{P(S)}{X} \leq \frac{1 - f(S)}{X}.$$

Hence,

$$P(\overline{A}) = 1 - P(A) \geq 1 - \frac{1 - f(S)}{X}$$

and, directly from the definition of $f(A)$,

$$P(\overline{A}) \geq f(A).$$

Using similar argumentation as for the first constraint, we obtain the second constraint:

$$f(A) \geq 1 - \frac{1 - f(S)}{X}.$$

Both constraints for the relation A **negates** S are the immediate consequences of their definitions:

$$t(A) = f(S)$$

and

$$f(A) = t(S).$$

The constraints for the relation A **disjoins** $\{S_1, S_2,..., S_n\}$ are derived
as follows. From

$$P(A) \geq P(S_i) \geq t(S_i)$$

we obtain the first constraint:

$$t(A) \geq t(S_i).$$

Furthermore,

$$P(\overline{S_i}) \geq P(\overline{A}) \geq f(A),$$

hence, the second constraint is

$$f(S_i) \geq f(A).$$

From

$$P(\overline{A}) = 1 - P(\bigcup_i S_i) \geq 1 - \sum_i P(S_i) \geq 1 - \sum_i (1 - f(S_i))$$

follows the next constraint:

$$f(A) \geq 1 - \sum_i (1 - f(S_i)) .$$

The last constraint is a consequence of

$$P(A) \leq \sum_i P(S_i).$$

Thus,

$$P(S_i) \geq P(A) - \sum_{j \neq i} P(S_j) \geq P(A) - \sum_{j \neq i} (1 - f(S_j))$$

and the constraint is

$$t(S_i) \geq t(A) - \sum_{j \neq i} (1 - f(S_j)).$$

The justification of the remaining constraints is similar to those given previously. Hence, it will be omitted.

A **disjoins-independent** $\{S_1, S_2,..., S_n\}$

$$t(A) \geq 1 - \prod_i (1 - t(S_i)),$$

$$f(A) \geq \prod_i f(S_i),$$

$$t(S_i) \geq 1 - \frac{1 - t(A)}{\prod_{j \neq i} f(S_j)},$$

$$f(S_i) \geq \frac{f(A)}{\prod_{j \neq i}(1 - t(S_j))},$$

A   disjoins-exclusive $\{S_1, S_2, ..., S_n\}$

$$t(A) \geq \sum_i t(S_i),$$

$$f(A) \geq 1 - \sum_i (1 - f(S_i)),$$

$$t(S_i) \geq t(A) - \sum_{j \neq i}(1 - f(S_j)),$$

$$f(S_i) \geq f(A) + \sum_{j \neq i} t(S_j),$$

A   conjoins $\{S_1, S_2, ..., S_n\}$

$$t(A) \geq 1 - \sum_i (1 - t(S_i)),$$

$$f(A) \geq f(S_i),$$

$$t(S_i) \geq t(A),$$

$$f(S_i) \geq f(A) - \sum_{j \neq i}(1 - t(S_j)),$$

A   conjoins-independent $\{S_1, S_2, ..., S_n\}$

$$t(A) \geq \prod_i t(S_i),$$

$$f(A) \geq 1 - \prod_i (1 - f(S_i)),$$

$$t(S_i) \geq \frac{t(A)}{\prod_{j \neq i}(1 - f(S_j))},$$

$$f(S_i) \geq 1 - \frac{1 - f(A)}{\prod_{j \neq i} t(S_j)}.$$

$\{S_1, S_2,..., S_n\}$ mutually exclusive

$$f(S_i) \geq \sum_{j \neq i} t(S_j).$$

5.2.5 Termination of Propagation

Suppose that for the inference net presented by Figure 5.8, $P(C)$ is found to be in the interval [0.6, 0.7]. Thus, $t(C) = 0.6$ and $f(C) = 0.3$. Initial values for all remaining bounds are equal to zero. Activated are the constraints for $f(A), f(B)$, and $f(E)$, and all three bounds will be set to 0.3.

Now say that the new information indicates that $P(B) \in [0.5, 0.55]$, or that $t(B) = 0.5$ and $f(B) = 0.45$. The former value for $f(B)$ does not restrict it anymore, since $0.3 < 0.45$. The following constraints are activated:

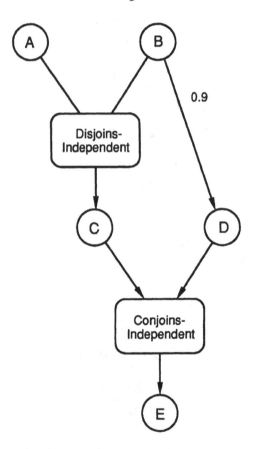

Figure 5.8 Inference Net

$$t(A) \geq 1 - \frac{0.4}{0.45} = 0.112,$$

$$f(A) \geq \frac{0.3}{1 - 0.5} = 0.6,$$

$$t(B) \geq 1 - \frac{0.4}{0.6} = 0.333,$$

$$f(B) \geq \frac{0.3}{1 - 0.112} = 0.338,$$

$$t(D) \geq 0.5 \cdot 0.9 = 0.45,$$

$$t(E) \geq 0.6 \cdot 0.45 = 0.27,$$

$$f(E) \geq 1 - 0.7 \cdot 1 = 0.3.$$

Therefore, $t(A) = 0.112, f(A) = 0.6, t(B) = 0.5, f(B) = 0.45, t(D) = 0.45, t(E) = 0.27$, and $f(E) = 0.3$.

Constraint for $t(C)$ is not activated because

$$1 - (1 - 0.112) \cdot (1 - 0.5) = 0.556,$$

and the current value for $t(C)$ is 0.6. Constraint for $f(C)$ is not activated because

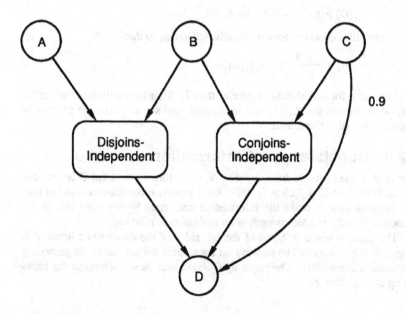

Figure 5.9 Inference Net

$$0.6 \cdot 0.45 = 0.27$$

and the current value for $f(C)$ is 0.3. Similarly, the second constraint for $t(D)$, namely

$$t(D) \geq \frac{t(E)}{1 - f(C)} = \frac{0.27}{1 - 0.3} = 0.386$$

is not activated because $t(D)$ has the value 0.45.

Another example of an inference net is given by Figure 5.9. Let us analyze it with the same assumptions about $P(C)$ as in the case of the inference net from Figure 5.8 (i.e., that $t(C) = 0.6$ and $f(C) = 0.3$). A number of constraints are activated, among them a constraint for $t(D)$, which follows from the fact that C enables D with strength 0.9, that is,

$$t(D) \geq 0.6 \cdot 0.9 = 0.54.$$

Another activated constraint is for $t(B)$, a consequence of the fact that D conjoins-independent $\{B, C\}$, that is,

$$t(B) \geq \frac{0.54}{1 - 0.3} = 0.771.$$

Yet another constraint is for $t(D)$, following from the fact that D disjoins-independent $\{A, B\}$, that is,

$$t(D) \geq 1 - (1 - 0.771) = 0.771.$$

In turn, the constraint for $t(B)$ is activated again, so that

$$t(B) \geq \frac{0.771}{1 - 0.3} = 1.101.$$

The value for $t(B)$ should be greater than 1.101, a contradiction. Nevertheless, the constraint for $t(D)$ is activated again, and so on. Thus the process of propagation will not terminate.

5.2.6 Consistency of Information

Note that in the example from Figure 5.9, the information about C is not consistent, because $t(C) + f(C) > 1$. INFERNO reports inconsistencies to the user and suggests how to make the information consistent by lowering one of the bounds or by reducing the strength of an enables-type relation.

The process needs analysis of the left sides of the constraints; hence it is propagation in reverse. The backing-up constraints are similar to the preceding propagation constraints. The paper by Quinlan lists these constraints for backing-up inconsistencies.

5.3 Concluding Remarks

The Dempster-Shafer theory has recently received much attention in the AI community as very promising for reasoning under uncertainty in expert systems. The Dempster-Shafer theory is not as demanding as probability theory (e.g., it does not need prior probabilities or conditional probabilities, and it permits the sum of belief for a proposition and belief for its negation to be smaller than one). Ways of handling uncertainty in MYCIN and PROSPECTOR, after some modifications, may be interpreted in fact as special cases of Dempster-Shafer theory (Gordon and Shortliffe, 1984; Grosof, 1986). However, the computational complexity of Dempster's rule of combination is enormous. A way of removing this obstacle by local computations was suggested (Shafer and Logan, 1987; Shenoy and Shafer, 1986). As a continuation of research in this direction, a scheme for propagating belief functions in qualitative Markov trees was created (Mellouli *et al.*, 1986; Shafer *et al.*, 1987, 1988).

Critics of the Dempster-Shafer theory argue that it is inadequate for empirical data (Lemmer, 1986) or that in some cases Dempster's rule of combination should not be applied at all (Zadeh, 1986b). In Dubois and Prade (1985) it was shown that assigning near zero values to basic probability numbers, compared with assigning zeros, may give completely different results by Dempster's rule of combination. In Kyburg (1986) the claim was made that closed convex sets of classical probability theory are more general than belief functions.

The INFERNO system has some advantages over PROSPECTOR, as pointed out in Quinlan (1983). Among the disadvantages of INFERNO is that the user is forced to make decisions about solving inconsistencies. An expert system should be able to reason on its own even when the evidence is contradictory (Mamdani *et al.*, 1985). The other obvious disadvantage is the problem of termination of the algorithm used in INFERNO. Also, INFERNO is too cautious. The bounds it produces are too wide (Pearl, 1988). Some improvements of INFERNO with respect to the problem of termination are suggested in Liu and Gammerman (1986). The new version allows more propagation in the network with few additional computations needed.

Exercises

1. Find a basic probability assignment $m: 2^\Theta \rightarrow [0, 1]$ and three subsets A, B, C of Θ such that

$$\text{Bel} (A \cup B) > \text{Bel} (A) + \text{Bel} (B) - \text{Bel} (A \cap B),$$

$$\text{Bel} (A \cup C) > \text{Bel} (A) + \text{Bel} (C) - \text{Bel} (A \cap C),$$

and

$$\text{Bel} (B \cup C) = \text{Bel} (B) + \text{Bel} (C) - \text{Bel} (B \cap C),$$

where

$$\text{Bel} (A \cap B) > 0, \text{Bel} (A \cap C) > 0, \text{ and } \text{Bel} (B \cap C) > 0.$$

2. For the function Bel: $2^\Theta \to [0, 1]$ find the basic probability assignment $m: 2^\Theta \to [0, 1]$ and the plausibility function Pl: $2^\Theta \to [0, 1]$, where $\Theta = \{0, 1, 2, 3\}$ and

$\text{Bel}(\{0\}) = \text{Bel}(\{1\}) = 0,$

$\text{Bel}(\{2\}) = \text{Bel}(\{3\}) = \text{Bel}(\{0, 2\}) = \text{Bel}(\{0, 3\}) = \text{Bel}(\{1, 2\})$

$\qquad = \text{Bel}(\{1, 3\}) = \frac{1}{4},$

$\text{Bel}(\{0,1\}) = \text{Bel}(\{2, 3\}) = \text{Bel}(\{0, 2, 3\}) = \text{Bel}(\{1, 2, 3\}) = \frac{1}{2},$

$\text{Bel}(\{0, 1, 2\}) = \text{Bel}(\{0, 1, 3\}) = \frac{3}{4}.$

3. For the function Bel: $2^\Theta \to [0, 1]$ find the basic probability assignment $m: 2^\Theta \to [0, 1]$ and the plausibility function Pl: $2^\Theta \to [0, 1]$, where $\Theta = \{1, 2, 3\}$ and

$\text{Bel}(\{1\}) = \text{Bel}(\{2\}) = 0,$

$\text{Bel}(\{3\}) = \text{Bel}(\{1, 3\}) = \frac{1}{2},$

$\text{Bel}(\{1, 2\}) = \frac{1}{4},$

$\text{Bel}(\{2, 3\}) = \frac{3}{4}.$

4. The frame Θ is $\{a, b, c\}$, and two basic probability assignments, m and n, are given here:

	$\{a\}$	$\{b\}$	$\{c\}$	$\{a, b\}$	$\{a, c\}$	$\{b, c\}$	$\{a, b, c\}$
m	0.3	0	0.2	0.3	0	0.1	0.1
n	0	0	0.2	0.2	0.3	0.2	0.1

Assuming independence of both pieces of evidence, tell their orthogonal sum $m \oplus n$. Draw the square as in Figures 5.3 and 5.4.

5. For the following inference network of INFERNO,

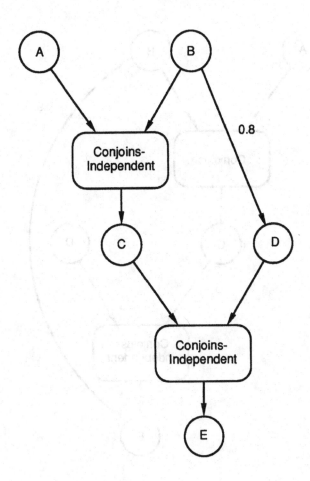

assume the following initial values: $t(A) = 0.5$, $f(A) = 0.4$, $t(B) = 0.6$, and $f(B)$ = 0.3. Find final values for $t(A), f(A)$, $t(B), f(B)$, $t(C), f(C)$, $t(D), f(D)$, $t(E)$, and $f(E)$.

6. For the following inference network of INFERNO,

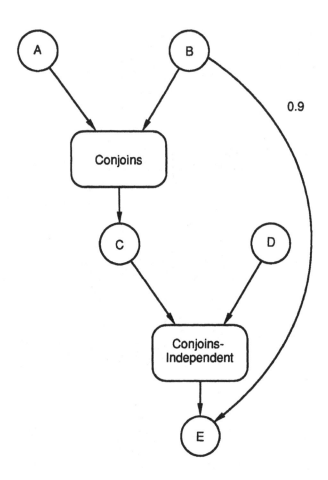

assume the following initial values: $t(A) = 0.5$, $f(A) = 0.4$, $t(B) = 0.6$, $f(B) = 0.3$, $t(D) = 0.7$, and $f(D) = 0.2$. Find final values for $t(A)$, $f(A)$, $t(B)$, $f(B)$, $t(C)$, $f(C)$, $t(D)$, $f(D)$, $t(E)$, and $f(E)$.

C H A P T E R
6

SET-VALUED
QUANTITATIVE
APPROACHES

At first glance, the three approaches to uncertainty presented in this chapter—
fuzzy set theory, incidence calculus, and rough set theory—have very little in
common. However, they are placed in the same chapter because uncertainty in
all three theories is characterized by a set. In the case of fuzzy set theory, a way
to represent uncertainty is a linguistic variable whose terms are characterized by
fuzzy subsets or a possibility distribution, which may also be viewed as a fuzzy
set. An incidence from incidence calculus is a set, a subset of the set of discourse. Finally, a rough set characterizes uncertainty by a family of subsets of
the universe.

6.1 Fuzzy Set Theory

The originator of fuzzy set theory is L. A. Zadeh (Zadeh, 1965). The theory has
been developed so extensively that currently an entire spectrum of fuzzy theories
has evolved. The theory of fuzzy sets was introduced to represent uncertainty,
especially the type of uncertainty that arises from imprecision and ambiguity, in
the sense of vagueness rather than incomplete information.

A number of expert systems have employed fuzzy set theory. Among them
are CADIAC-2, an expert system for medical diagnosis created by K.-P. Adlassnig and G. Kolarz (Adlassnig, 1982; Adlassnig and Kolarz, 1982) and SPERIL
I, an expert system for assessing structural damage to buildings after earthquake,
by M. Ishizuka, K. S. Fu, and J. T. P. Yao (Ishizuka *et al.*, 1982). A VLSI

chip for inference in fuzzy logic was developed by AT&T Bell Laboratories (Togai and Watanabe, 1986).

6.1.1 Fuzzy Sets and Operations

The basic concept is that of a *fuzzy subset* of the *universe of discourse* U, a subset A of U, where each element x of U is characterized by the value of a function, called the *membership function* μ_A. Usually the fuzzy subset A on U is just called a fuzzy set A. Value $\mu_A(u)$, for $u \in U$, is a number from the real interval [0, 1], called a *grade of membership* of u. Thus, A is a subset of U with no sharp boundary. Therefore, a fuzzy set is characterized by the membership function $\mu_A: U \rightarrow [0, 1]$. A special case of a fuzzy set is an ordinary (crisp) set, where $\mu_A: U \rightarrow \{0, 1\}$ is called a *characteristic function*, $\mu_A(u) = 1$ if $x \in A$ and $\mu_A(u) = 0$ otherwise.

Two fuzzy sets A and B are equal, denoted $A = B$, if and only if for all $u \in U$

$$\mu_A(u) = \mu_B(u).$$

The empty set $\emptyset$ is defined as follows: For all $u \in U$

$$\mu_\emptyset(u) = 0.$$

For the universe U,

Table 6.1 Fuzzy Set FAST

u	$\mu_{FAST}(u)$
0	0
10	0.01
20	0.02
30	0.05
40	0.1
50	0.4
60	0.8
70	0.9
80	1

$$\mu_U(u) = 1$$

for any $u \in U$.

An example of the fuzzy set FAST is presented in Table 6.1. The universe U of FAST is a set $\{0, 10, 20, 30, 40, 50, 60, 70, 80\}$ of possible values of a speed of a car in miles per hour. The membership function μ_{FAST} represents judgment or perception of an expert. Thus, each member of U is qualified as being a member of the set FAST with a grade. Low speeds have been assigned low values. In the example, speed 0 miles per hour has been assigned grade 0— a parked car is not fast. On the other hand, according to an expert, a car driven at the speed of 80 miles per hour is fast. It should be emphasized that function μ_{FAST} has its values assigned in an arbitrary way, and different experts may come up with different values. The fuzzy set FAST may be represented by showing its grade of membership, that is, FAST $=\{(0, 0), (10, 0.01), (20, 0.02), (30, 0.05), (40, 0.1), (50, 0.4), (60, 0.8), (70, 0.9), (80, 1)\}$.

In this section, terms will be distinguished from their denotations by using uppercase letters for the latter. For example, the term *fast car* has its denotation as a fuzzy subset FAST.

Another example of a fuzzy set, the set DANGEROUS, is presented in Table 6.2. The universe U of DANGEROUS is the same set as for the fuzzy set FAST. Two more examples of fuzzy sets are presented in Figures 6.3 and 6.4. Figure 6.1 presents fuzzy set AROUND_FIVE $= \{(u, \mu_{AROUND\_FIVE}(u)) \mid u \in [0, 10]\}$, where

Table 6.2 Fuzzy Set DANGEROUS

u	$\mu_{DANGEROUS}(u)$
0	0
10	0.05
20	0.1
30	0.15
40	0.2
50	0.3
60	0.7
70	1
80	1

$$\mu_{\text{AROUND\_FIVE}}(u) = \begin{cases} 0 & \text{if } 0 \le u < 2, \\[2ex] \dfrac{u - 2}{2} & \text{if } 2 \le u < 4, \\[2ex] 1 & \text{if } 4 \le u \le 6, \\[2ex] \dfrac{8 - u}{2} & \text{if } 6 \le u < 8, \\[2ex] 0 & \text{if } 8 \le u \le 10. \end{cases}$$

Figure 6.2 presents fuzzy set BIG_NUMBER = $\{(u, \mu_{\text{BIG\_NUMBER}}(u) \mid u \in \mathbb{R}\}$, where

$$\mu_{\text{BIG\_NUMBER}}(u) = \begin{cases} 0 & \text{if } u \le 1, \\[2ex] \dfrac{(u - 1)^2}{u^2 + u} & \text{if } u > 1. \end{cases}$$

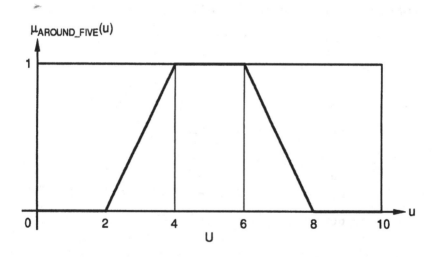

Figure 6.1 Graph of Fuzzy Set AROUND_FIVE

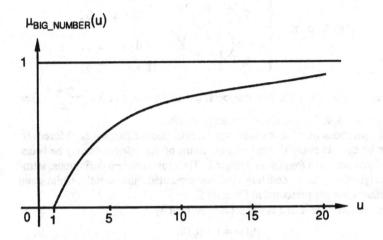

Figure 6.2 Graph of Fuzzy Set BIG_NUMBER

Some forms of fuzzy sets are well known and broadly accepted. One of them is a fuzzy set whose membership function is a standard function called *S-function*, defined on the universe of discourse $U = [a, b]$ as follows:

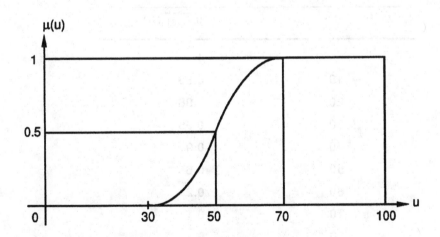

Figure 6.3 Graph of Fuzzy Set Defined by S-Function

$$S(u; \alpha, \beta, \gamma) = \begin{cases} 0 & \text{if } a \leq u \leq \alpha, \\ 2\left(\dfrac{u - \alpha}{\gamma - \alpha}\right)^2 & \text{if } \alpha \leq u \leq \beta, \\ 1 - 2\left(\dfrac{u - \gamma}{\gamma - \alpha}\right)^2 & \text{if } \beta \leq u \leq \gamma, \\ 1 & \text{if } \gamma < u \leq b, \end{cases}$$

where a, b, α, β, and γ are real numbers, $a \leq \alpha \leq \gamma \leq b$, and $\beta = \dfrac{\alpha + \gamma}{2}$ (see Figure 6.3 for $S(u; 30, 50, 70)$, $a = 0$, and $b = 100$).

All operations on ordinary sets may be extended for fuzzy sets. Moreover, this can be done in many different ways. Some of the extensions may be better justified and are more frequently accepted. The most common definitions, introduced originally by L. A. Zadeh in 1965, are presented subsequently, while some of the other ways are presented in Exercise 5.

The *complement* $\overline{A}$ of a fuzzy set A is defined by

$$\mu_{\overline{A}}(u) = 1 - \mu_A(u)$$

for all $u \in A$. The complement of the fuzzy set FAST is presented in Table 6.3.

In Bellman and Gertz (1973) the preceding definition of complementation was justified as the only function f that depends only on μ_A, $f(0) = 1$ and $f(1) = 0$, and such that

Table 6.3 Fuzzy Set $\overline{\text{FAST}}$

u	$\mu_{\overline{\text{FAST}}}(u)$
0	1
10	0.99
20	0.98
30	0.95
40	0.9
50	0.6
60	0.2
70	0.1
80	0

$$\mu_A(u_1) - \mu_A(u_2) = f(\mu_A(u_2)) - f(\mu_A(u_1)).$$

Let A and B be fuzzy sets over the same universe U. The *union of* A *and* B, denoted $A \cup B$, is defined as follows:

$$\mu_{A \cup B}(u) = \max(\mu_A(u), \mu_B(u))$$

for all $u \in U$. The union of the fuzzy sets FAST and DANGEROUS is presented in Table 6.4.

The *intersection of* A *and* B, denoted $A \cap B$, is defined as follows

$$\mu_{A \cap B}(u) = \min(\mu_A(u), \mu_B(u))$$

for all $u \in U$. The intersection of the fuzzy set FAST and DANGEROUS is presented in Table 6.5.

In Bellman and Gertz (1973) the preceding definitions of union and intersection were justified as the only functions f and g, respectively, that

1. Depend only on μ_A and μ_B,

2. Are commutative, that is,

$$f(\mu_A(u), \mu_B(u)) = f(\mu_B(u), \mu_A(u)) \text{ and } g(\mu_A(u), \mu_B(u)) = g(\mu_B(u), \mu_A(u))$$

for all $u \in U$,

3. Are associative, that is,

Table 6.4 Fuzzy Set FAST ∪ DANGEROUS

u	$\mu_{\text{FAST} \cup \text{DANGEROUS}}(u)$
0	0
10	0.05
20	0.1
30	0.15
40	0.2
50	0.4
60	0.8
70	1
80	1

$$f(f(\mu_A(u), \mu_B(u)), \mu_C(u)) = f(\mu_A(u), f(\mu_B(u), \mu_C(u)))$$

and

$$g(g(\mu_A(u), \mu_B(u)), \mu_C(u)) = g(\mu_A(u), g(\mu_B(u), \mu_C(u)))$$

for all $u \in U$,

    4. Are distributive over the other, that is,

$$f(\mu_A(u), g(\mu_B(u), \mu_C(u))) = g(f(\mu_A(u), \mu_B(u)), f(\mu_A(u), \mu_C(u)))$$

and

$$g(\mu_A(u), f(\mu_B(u), \mu_C(u))) = f(g(\mu_A(u), \mu_B(u)), g(\mu_A(u), \mu_C(u)))$$

for all $u \in U$,

    5. Are continuous, that is, a small increase of $\mu_A(u)$ or $\mu_B(u)$ implies a small increase of $f(\mu_A(u), \mu_B(u))$ and $g(\mu_A(u), \mu_B(u))$,

    6. Are nondecreasing, that is,

$$\mu_A(u_1) \geq \mu_A(u_2) \text{ and } \mu_B(u_1) \geq \mu_B(u_2) \text{ implies}$$

$$f(\mu_A(u_1), \mu_B(u_1)) \geq f(\mu_A(u_2), \mu_B(u_2)) \text{ and } g(\mu_A(u_1), \mu_B(u_1)) \geq g(\mu_A(u_2), \mu_B(u_2))$$

for all $u_1, u_2 \in U$,

Table 6.5 Fuzzy Set FAST ∩ DANGEROUS

u	$\mu_{\text{FAST} \cap \text{DANGEROUS}}(u)$
0	0
10	0.01
20	0.02
30	0.05
40	0.1
50	0.3
60	0.7
70	0.9
80	1

7. $f(\mu, \mu)$ and $g(\mu, \mu)$ are strictly increasing, that is,

$$\mu_A(u_1) = \mu_B(u_1) > \mu_A(u_2) = \mu_B(u_2) \quad \text{implies}$$

$f(\mu_A(u_1), \mu_B(u_1)) > f(\mu_A(u_2), \mu_B(u_2))$ and $g(\mu_A(u_1), \mu_B(u_1)) > g(\mu_A(u_2), \mu_B(u_2))$

for all $u_1, u_2 \in U$,

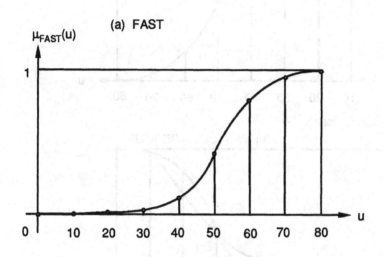

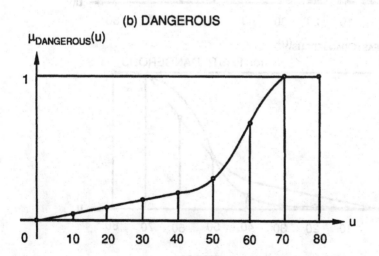

Figure 6.4 Graphs of Fuzzy Sets FAST and DANGEROUS

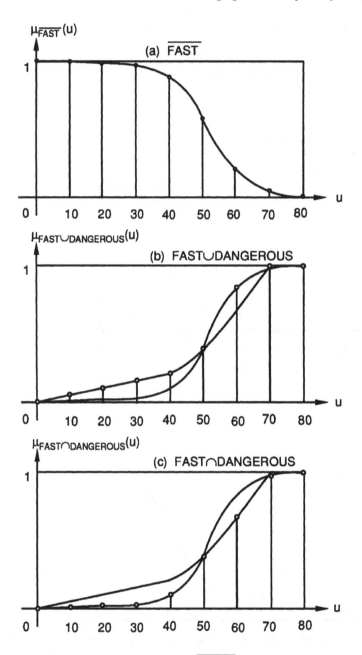

Figure 6.5 Graphs of Fuzzy Sets $\overline{\text{FAST}}$, FAST∪DANGEROUS, and FAST∩DANGEROUS

8. Membership in $A \cup B$ requires less than membership in A or B and membership in $A \cap B$ requires more than membership in A or B, that is,

$$\max (\mu_A(u), \mu_B(u)) \leq f(\mu_A(u), \mu_B(u)) \text{ and } \min (\mu_A(u), \mu_B(u)) \geq g(\mu_A(u), \mu_B(u))$$

for all $u \in U$, and finally,

9. u certainly not being in A and B implies u certainly not being in $A \cup B$, and u certainly being in A and B implies u certainly being in $A \cap B$, that is,

$$\mu_A(u) = 0 \text{ and } \mu_B(u) = 0 \text{ implies } f(\mu_A(u), \mu_B(u)) = 0$$

and

$$\mu_A(u) = 1 \text{ and } \mu_B(u) = 1 \text{ implies } g(\mu_A(u), \mu_B(u)) = 1,$$

for all $u \in U$.

6.1.2 Extended Venn Diagrams

In general, fuzzy sets cannot be represented by Venn diagrams. Instead, their graphs may be used, as originally suggested by L. A. Zadeh. The idea is presented in Figures 6.4 and 6.5. For a better view, the fuzzy sets in the figures are

Table 6.6 Fuzzy Relation GREATER_THAN

	U_2								
	0	10	20	30	40	50	60	70	80
0	0	0.1	0.2	0.3	0.4	0.5	0.7	0.9	1
10	0	0	0.1	0.2	0.3	0.4	0.5	0.7	0.9
20	0	0	0	0.1	0.2	0.3	0.4	0.5	0.7
30	0	0	0	0	0.1	0.2	0.3	0.4	0.5
U_1 40	0	0	0	0	0	0.1	0.2	0.3	0.4
50	0	0	0	0	0	0	0.1	0.2	0.3
60	0	0	0	0	0	0	0	0.1	0.2
70	0	0	0	0	0	0	0	0	0.1
80	0	0	0	0	0	0	0	0	0

drawn as continuous functions, in spite of the fact that they should be represented by separated points. The interpretation for all cases is straightforward from the definitions.

6.1.3 Fuzzy Relations and Operations

The concept of a fuzzy relation was introduced by L. A. Zadeh in 1965. Let U_1, U_2,..., U_n be universes of discourse. Any fuzzy set on $U_1 \times U_2 \times \cdots \times U_n$ is called an n-ary fuzzy relation R on $U_1 \times U_2 \times \cdots \times U_n$, or briefly, a fuzzy relation R. Thus,

$$R = \{((u_1, u_2,..., u_n), \mu_R(u_1, u_2,..., u_n)) \mid u_i \in U_i, i = 1, 2,..., n\}.$$

An example of the binary fuzzy relation GREATER_THAN is presented in Table 6.6.

Let $s = (i_1, i_2,..., i_k)$ be a subsequence of $(1, 2,..., n)$ and let $\bar{s} = (i_{k+1}, i_{k+2},..., i_n)$ be the sequence complementary to $(i_1, i_2,..., i_k)$, where $1 \leq k \leq n - 1$.

For example, $s = (1, 4, 5)$ is a subsequence of $(1, 2, 3, 4, 5)$, and $\bar{s} = (2, 3)$. The *projection of an* n-*ary fuzzy relation* R on $U_{(s)} = U_{i_1} \times U_{i_2} \times \cdots \times U_{i_k}$,

Table 6.7 The Cylindrical Extension c(FAST1) of FAST1 in $U_1 \times U_2$

				U_2					
	0	10	20	30	40	50	60	70	80
0	0	0	0	0	0	0	0	0	0
10	0.01	0.01	0.01	0.01	0.01	0.01	0.01	0.01	0.01
20	0.02	0.02	0.02	0.02	0.02	0.02	0.02	0.02	0.02
30	0.05	0.05	0.05	0.05	0.05	0.05	0.05	0.05	0.05
U_1 40	0.1	0.1	0.1	0.1	0.1	0.1	0.1	0.1	0.1
50	0.4	0.4	0.4	0.4	0.4	0.4	0.4	0.4	0.4
60	0.8	0.8	0.8	0.8	0.8	0.8	0.8	0.8	0.8
70	0.9	0.9	0.9	0.9	0.9	0.9	0.9	0.9	0.9
80	1	1	1	1	1	1	1	1	1

denoted $\text{Proj}_{U_{(s)}}(R)$, is a k-ary fuzzy relation:

$$\{((u_{i_1}, u_{i_2},..., u_{i_k}), \sup_{u_{i_{k+1}}, u_{i_{k+2}},..., u_{i_n}} \mu_R(u_1, u_2,...,u_n)) \mid (u_{i_1}, u_{i_2},..., u_{i_k}) \in U_{(s)}\}.$$

In the preceding formula, *sup* means the least upper bound. When universes $U_1, U_2, .., U_n$ are finite, *sup* may be replaced by *max*. For $k = 1$, k-ary fuzzy relations are ordinary fuzzy sets on a single universe.

For example,

$\text{Proj}_{U_1}(\text{GREATER\_THAN})$

$= \{(u, \max_v \mu_{\text{GREATER\_THAN}}(u, v)) \mid u \in U_1\}$

$= \{(0, 0), (10, 0.1), (20, 0.2), (30, 0.3), (40, 0.4), (50, 0.5), (60, 0.7), (70, 0.9), (80, 1)\}$,

$\text{Proj}_{U_2}(\text{GREATER\_THAN})$

$= \{(v, \max_u \mu_{\text{GREATER\_THAN}}(u, v)) \mid v \in U_2\}$

Table 6.8 The Cylindrical Extension c(FAST2) of FAST2 in $U_1 \times U_2$

					U_2				
	0	10	20	30	40	50	60	70	80
0	0	0.01	0.02	0.05	0.1	0.4	0.8	0.9	1
10	0	0.01	0.02	0.05	0.1	0.4	0.8	0.9	1
20	0	0.01	0.02	0.05	0.1	0.4	0.8	0.9	1
30	0	0.01	0.02	0.05	0.1	0.4	0.8	0.9	1
U_1 40	0	0.01	0.02	0.05	0.1	0.4	0.8	0.9	1
50	0	0.01	0.02	0.05	0.1	0.4	0.8	0.9	1
60	0	0.01	0.02	0.05	0.1	0.4	0.8	0.9	1
70	0	0.01	0.02	0.05	0.1	0.4	0.8	0.9	1
80	0	0.01	0.02	0.05	0.1	0.4	0.8	0.9	1

$= \{(0, 1), (10, 0.9), (20, 0.7), (30, 0.5), (40, 0.4), (50, 0.3), (60, 0.2), (70, 0.1), (80, 0)\}$,

The converse of the projection of n-ary relation is called a *cylindrical extension*. Let R be a k-ary fuzzy relation on $U_{(s)} = U_{i_1} \times U_{i_2} \times \cdots \times U_{i_k}$. A cylindrical extension of R in $U = U_1 \times U_2 \times \cdots \times U_n$ is an n-ary relation, denoted $c(R)$ and defined as follows:

$$c(R) = \{((u_1, u_2,..., u_n), \mu_R(u_{i_1}, u_{i_2},..., u_{i_k}) \mid (u_1, u_2,..., u_n) \in U\}.$$

For example, let FAST1 and FAST2 be two fuzzy sets on universes U_1 and U_2, respectively, where $U_1 = U_2 = U = \{0, 10, 20, 30, 40, 50, 60, 70, 80\}$, and

$$\mu_{FAST1}(u) = \mu_{FAST2}(u) = \mu_{FAST}(u)$$

for all $u \in U$, where μ_{FAST} is defined by Table 6.1. The cylindrical extension of FAST1 in $U_1 \times U_2$ is presented in Table 6.7, while the cylindrical extension of FAST2 in $U_1 \times U_2$ is presented in Table 6.8

Let R be an r-ary fuzzy relation on $U_1 \times U_2 \times \cdots \times U_r$ and S be an $(n - s +1)$-ary fuzzy relation on $U_s \times U_{s+1} \times \cdots \times U_n$ where $1 \le s \le r \le n$. The *join of* R *and* S is defined as

Table 6.9 The Join of c(FAST1) and c(FAST2)

		U_2							
	0	10	20	30	40	50	60	70	80
0	0	0	0	0	0	0	0	0	0
10	0	0.01	0.01	0.01	0.01	0.01	0.01	0.01	0.01
20	0	0.01	0.02	0.02	0.02	0.02	0.02	0.02	0.02
30	0	0.01	0.02	0.05	0.05	0.05	0.05	0.05	0.05
U_1 40	0	0.01	0.02	0.05	0.1	0.1	0.1	0.1	0.01
50	0	0.01	0.02	0.05	0.1	0.4	0.4	0.4	0.04
60	0	0.01	0.02	0.05	0.1	0.4	0.8	0.8	0.08
70	0	0.01	0.02	0.05	0.1	0.4	0.8	0.9	0.9
80	0	0.01	0.02	0.05	0.1	0.4	0.8	0.9	1

$$c(R) \cap c(S)$$

where $c(R)$ and $c(S)$ are cylindrical extensions of R and S, respectively, on $U_1 \times U_2 \times \cdots \times U_n$.

The join of $c(\text{FAST1})$ and $c(\text{FAST2})$ is presented in Table 6.9.

Fuzzy relations may be combined with each other by different versions of composition. Here a sup-min composition is cited; other definitions are cited in Exercise 8.

Let R be an r-ary fuzzy relation on $U_1 \times U_2 \times \cdots \times U_r$ and S be an $(n - s + 1)$-ary fuzzy relation on $U_s \times U_{s+1} \times \cdots \times U_n$, where $1 \leq s \leq n$. Let $(\{1, 2,..., r\} - \{s, s + 1,..., n\}) \cup (\{s, s+1,..., n\} - \{1, 2,..., r\})$ be denoted $\{i_1, i_2,..., i_k\}$. The set $\{i_1, i_2,..., i_k\}$ is the symmetric difference of $\{1, 2,..., r\}$ and $\{s, s + 1,..., n\}$. The *composition of* R *and* S, denoted $R \circ S$, is the following fuzzy relation:

$$\text{Proj}_{(U_{i_1}, U_{i_2},..., U_{i_k})}(c(R) \cap c(S))$$

that is, the projection of the join of $C(R)$ and $C(S)$ on $U_{i_1} \times U_{i_2} \times \cdots \times U_{i_k}$.

Two special cases are of greater importance:

1. $r = 1 = s$ and $n = 2$. Then the composition $R \circ S$ of R and S may be computed by the following formula:

Table 6.10 The Composition of FAST and GREATER_THAN

u	$\mu_{\text{FAST}} \circ \text{GREATER\_THAN}(u)$
0	0
10	0
20	0.01
30	0.02
40	0.05
50	0.1
60	0.1
70	0.2
80	0.3

$$R \circ S = \{(x_2, \sup_{u_1} \min(\mu_R(u_1), \mu_S(u_1, u_2))) \mid u_1 \in U_1, u_2 \in U_2\}.$$

2. $r = 2 = s$ and $n = 3$. In this case $R \circ S$ is given by

$$R \circ S = \{((u_1, u_3), \sup_{u_2} \min(\mu_R(u_1, u_2), \mu_S(u_2, u_3))) \mid u_1 \in U_1, u_2 \in U_2, u_3 \in U_3\}.$$

Examples of FAST $\circ$ GREATER_THAN and GREATER_THAN $\circ$ GREATER_THAN are given in Tables 6.10 and 6.11, respectively. All computations follow directly from definitions. For example,

$$\begin{aligned}\mu_{FAST \circ GREATER_THAN}(80) &= \max_{u_1} \min(\mu_{FAST}(u_1), \mu_{GREATER_THAN}(u_1, 80)) \\ &= \max(\min(\mu_{FAST}(0), \mu_{GREATER_THAN}(0, 80)), \\ &(\min(\mu_{FAST}(10), \mu_{GREATER_THAN}(10, 80)), \\ &(\min(\mu_{FAST}(20), \mu_{GREATER_THAN}(20, 80)), \\ &(\min(\mu_{FAST}(30), \mu_{GREATER_THAN}(30, 80)), \\ &(\min(\mu_{FAST}(40), \mu_{GREATER_THAN}(40, 80)), \\ &(\min(\mu_{FAST}(50), \mu_{GREATER_THAN}(50, 80)),\end{aligned}$$

Table 6.11 The Composition of GREATER_THAN and GREATER_THAN

		U_2								
		0	10	20	30	40	50	60	70	80
	0	0	0	0.1	0.1	0.2	0.2	0.2	0.2	0.4
	10	0	0	0	0.1	0.1	0.2	0.2	0.3	0.3
	20	0	0	0	0	0.1	0.1	0.2	0.2	0.3
	30	0	0	0	0	0	0.1	0.1	0.2	0.2
U_1	40	0	0	0	0	0	0	0.1	0.1	0.1
	50	0	0	0	0	0	0	0	0.1	0.1
	60	0	0	0	0	0	0	0	0	0.1
	70	0	0	0	0	0	0	0	0	0
	80	0	0	0	0	0	0	0	0	0

$$\text{(min } (\mu_{FAST}(60), \mu_{GREATER\_THAN}(60, 80),$$
$$\text{(min } (\mu_{FAST}(70), \mu_{GREATER\_THAN}(70, 80),$$
$$\text{(min } (\mu_{FAST}(80), \mu_{GREATER\_THAN}(80, 80))$$

$$\begin{aligned}
= \text{max } &(\text{min } (0, 0), \\
&\text{min } (0.01, 0.9), \\
&\text{min } (0.02, 0.7), \\
&\text{min } (0.05, 0.5), \\
&\text{min } (0.1, 0.4), \\
&\text{min } (0.4, 0.3), \\
&\text{min } (0.8, 0.2), \\
&\text{min } (0.9, 0.1), \\
&\text{min } (1, 0))
\end{aligned}$$

$$= \text{max } (0, 0.01, 0.02, 0.05, 0.1, 0.3, 0.2, 0.1, 0)$$
$$= 0.3.$$

6.1.4 Possibility Theory and Possibility Distributions

The theory of possibility was developed by L. A. Zadeh (1978) (see also Zadeh, 1979, 1981). One of the main concepts of possibility theory is that of a possibility distribution. Let $X = (X_1, X_2,..., X_n)$ be a variable on $U = U_1 \times U_2 \times \cdots \times U_n$. Let R be a fuzzy relation on U. Relation R may be interpreted as a constraint on the values that may be assigned to X. In other words, R induces a *possibility distribution* Π associated with X. That possibility distribution is described by a *possibility distribution function* $\pi: U \rightarrow [0, 1]$ which assigns a degree of possibility to every value u of X. Thus,

$$\Pi(X = u) = \pi(u) = \mu_R(u).$$

In an example of the fuzzy set FAST, a possibility distribution Π, associated with a speed X, is described as follows:

$$\Pi(X = 0) = 0,$$

$$\Pi(X = 10) = 0.01,$$

$$\Pi(X = 20) = 0.02,$$

$$\Pi(X = 30) = 0.05,$$

$$\Pi(X = 40) = 0.1,$$

$$\Pi(X = 50) = 0.4,$$

$$\Pi(X = 60) = 0.8,$$

$$\Pi(X = 70) = 0.9,$$

$\Pi(X = 80) = 1.$

The first equation just listed may be interpreted as follows: The possibility that a car with speed X equal to zero is qualified as fast is zero. Similarly, the second equation states that the possibility that a car with speed X equal to 10 mph is fast is 0.01, and so on.

As D. Dubois and H. Prade observed (1980), a fuzzy set R may be described by a fuzzy value that is assigned to a variable X. A possibility restriction R is the fuzzy set of nonfuzzy values that can possibly be assigned to X.

6.1.5 Linguistic Variables, Linguistic Modifiers, and the Translation Modifier Rule

Variables whose values are words or sentences in natural or artificial languages are called *linguistic*. Values of a linguistic variable are called *terms*. A fuzzy set is associated with each term over the universe of discourse U. The *base variable* is defined over U. An example of a linguistic variable *Body temperature* is presented in Figure 6.6. In the example, the terms are *low, normal, subfebrile, high,* and *very high*. The base variable is a temperature in degrees of Fahrenheit.

Let X be an element of universe U. Let A be a fuzzy subset of U. Then "X is A" is a proposition. For example, if U is the set $\{0, 10,..., 80\}$ and X is 70, then "70 is the speed of a fast car" (i.e., 70 belongs to the fuzzy subset FAST over the universe U of car speeds). Inference with propositions like "X is A" is performed via translation rules, to translate propositions into possibility distributions. The first such rule is called a *modifier rule*. Suppose "X is A" is represented by a possibility distribution function π, equal to μ_A. The question is

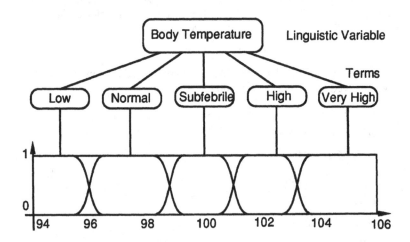

Figure 6.6 Linguistic Variable *Body temperature*

what is a possibility distribution representing "X is mA", where m is a *linguistic modifier* "not", "very", "fairly", and so on. The answer is given by a function f, where μ_{mA} is a composition of f and μ_A. Functions f are chosen arbitrarily. Usually, "not" is represented by $f(x) = 1 - x$, "very" by $f(x) = x^2$, "fairly" by $f(x) = x^{1/2}$, and so on.

Table 6.12 presents the fuzzy subset VERY_FAST, a modified fuzzy subset FAST from Table 6.1.

One of the important and frequently applied linguistic variables is *Truth*. There are different definitions of *Truth*—the one cited here is one introduced by J. F. Baldwin (1979a). Let U be the set $[0, 1]$. Then the term "true" of the linguistic variable *Truth* is defined as follows

$$\text{true} = \{(u, u) \mid u \in U\}$$

(see Figure 6.7). "False" is interpreted as "not true", that is, a complement of "true". The terms "absolutely true" and "absolutely false" are defined as follows:

$$\text{Absolutely true} = \{(u, \mu(u)) \mid \mu(u) = 0 \text{ if } u \neq 1, \mu(1) = 1\},$$

$$\text{Absolutely false} = \{(u, \mu(u)) \mid \mu(u) = 0 \text{ if } u \neq 0, \mu(0) = 1\}.$$

The linguistic modifiers "very" and "fairly" are used to represent the terms "very true", "fairly true", "very false", and "fairly false".

Table 6.12 Fuzzy Set VERY_FAST

u	$\mu_{\text{VERY\_FAST}}(u)$
0	0
10	0.0001
20	0.0004
30	0.0025
40	0.01
50	0.16
60	0.64
60	0.81
80	1

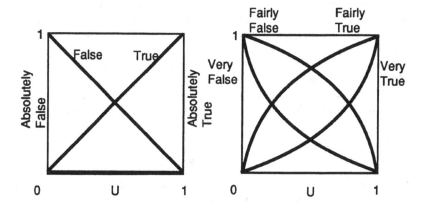

Figure 6.7 Terms of the Linguistic Variable *Truth*

6.1.6 Fuzzy Logic: A PRUF Approach

Although the concept of fuzzy logic is not unique and there is no system universally accepted, common ideas for fuzzy logics follow from the infinitely many-valued logic of J. Lukasiewicz. In classical logic, a proposition *p* is true or false. In fuzzy logics, the truth value of *p* may be a term of the linguistic variable "truth". To be more specific, propositions may be quantified by *fuzzy quantifiers*, such as "usually", "frequently", "seldom", "few", "most", and so on. Propositions may be qualified by the following concepts of fuzzy logic: *truth, probability,* and *possibility*. The system of fuzzy logic described here is based on Zadeh's approach, known as the meaning representation language PRUF (Zadeh, 1978). PRUF stands for Possibilistic Relational Universal Fuzzy. There are two main components of PRUF: translation rules and inference rules. A translation rule is used for assigning the possibility distribution to a proposition.

6.1.6.1 Propositions of Fuzzy Logic

The following are examples of propositions in fuzzy logic: "Econobox is a fast car" is a proposition, neither quantified nor qualified. The typical form of this proposition is "*X* is *A*", where *X* is a variable over the universe *U* and *A* is a fuzzy subset of *U*.

"Expensette is a very fast car" is a modified proposition, neither quantified nor qualified. It is modified by a modifier "very". The typical form of this proposition is "*X* is *mA*" (see 6.1.5).

"Econobox is a fast car is fairly true" is a qualified proposition by a fuzzy truth term "fairly true". "Econobox is a fast car is likely" or "Econobox is likely a fast car" are propositions qualified by a fuzzy probability "likely". "Econobox is a fast car is quite possible" is a qualified proposition by a fuzzy possibility "quite possible". The typical form is "X is A is λ", where X is a variable over the universe U, A is a fuzzy subset of U, and λ is a fuzzy truth term, fuzzy probability, or fuzzy possibility.

"If a car is fast then it is dangerous" is a conditional proposition, neither quantified nor qualified. The typical form of this proposition is "If X is A then Y is B", where X and Y are variables over the universe U and A and B are fuzzy subsets of U.

"Most fast cars are dangerous" is a conditional, qualified proposition, quantified by a fuzzy quantifier "Most". The typical form is "Q As are Bs", where Q is a fuzzy quantifier and A and B are fuzzy subsets of U.

"If a car is fast then it is likely dangerous", or "If a car is fast then it is dangerous is likely" are conditional, qualified propositions, qualified by a fuzzy probability "likely". The typical form is "If X is A then Y is B is λ", where X and Y are variables over the universe U, A and B are fuzzy subsets of U, and λ is a fuzzy probability.

Table 6.13 "X is fast or Y is dangerous"

U_2

	0	10	20	30	40	50	60	70	80
0	0	0.05	0.1	0.15	0.2	0.3	0.7	1	1
10	0.01	0.05	0.1	0.15	0.2	0.3	0.7	1	1
20	0.02	0.05	0.1	0.15	0.2	0.3	0.7	1	1
30	0.05	0.05	0.1	0.15	0.2	0.3	0.7	1	1
U_1 40	0.1	0.1	0.1	0.15	0.2	0.3	0.7	1	1
50	0.4	0.4	0.4	0.4	0.4	0.4	0.7	1	1
60	0.8	0.8	0.8	0.8	0.8	0.8	0.8	1	1
70	0.9	0.9	0.9	0.9	0.9	0.9	0.9	1	1
80	1	1	1	1	1	1	1	1	1

6.1.6.2   Translation Rules

There are four different translational rules in PRUF:

1. *Modifier rules* (see 6.1.5).

2. *Composition rules.* Composition rules translate a proposition that is a composition of proposition p and q into a possibility distribution function.

Let $X = (X_1, X_2,..., X_r)$ be a variable on the universe $U = U_1 \times U_2 \times \cdots \times U_r$ and let $Y = (Y_s, Y_{s+1},..., Y_n)$ be a variable on universe $V = U_1 \times U_2 \times \cdots \times U_n$, where $1 \le s \le r \le n$. Let A and B be fuzzy subsets of U and V, respectively. Then, "X is A or Y is B", "X is A and Y is B", and the conditional proposition "If X is A then Y is B" are translated into the possibility distribution function $\pi(X, Y)$, equal to the membership function of the following fuzzy sets:

$$c(A) \cup c(B),$$
$$c(A) \cap c(B),$$

and

$$c(\overline{A}) \oplus c(B)$$

Table 6.14 "X is fast and Y is dangerous"

U_2

	0	10	20	30	40	50	60	70	80
0	0	0	0	0	0	0	0	0	0
10	0	0.01	0.01	0.01	0.01	0.01	0.01	0.01	0.01
20	0	0.02	0.02	0.02	0.02	0.02	0.02	0.02	0.02
30	0	0.05	0.05	0.05	0.05	0.05	0.05	0.05	0.05
U_1 40	0	0.05	0.1	0.1	0.1	0.1	0.1	0.1	0.1
50	0	0.05	0.1	0.15	0.2	0.3	0.4	0.4	0.4
60	0	0.05	0.1	0.15	0.2	0.3	0.7	0.8	0.8
70	0	0.05	0.1	0.15	0.2	0.3	0.7	0.9	0.9
80	0	0.05	0.1	0.15	0.2	0.3	0.7	1	1

respectively, where $\overline{A}$ denotes the complement of A and $\oplus$ is the bounded sum from Exercise 5.

For a special case $r = 1$ and $s = n = 2$, the above fuzzy sets are defined as follows

$$c(A) \cup c(B) = \{((u, v), \max\ (\mu_A(u), \mu_B(v)))\ |\ (u, v) \in U \times V\},$$

$$c(A) \cap c(B) = \{((u, v), \min\ (\mu_A(u), \mu_B(v)))\ |\ (u, v) \in U \times V\},$$

$$c(\overline{A}) \oplus c(B) = \{((u, v), \min\ (1, 1 - \mu_A(u) + \mu_B(v)))\ |\ (u, v) \in U \times V\}.$$

For example, the proposition "X is fast or Y is dangerous", is translated into the possibility distribution function $\pi_{(X, Y)}$, presented in Table 6.13. This may be interpreted as follows. The possibility that a car with the speed X is fast or that a car with the speed Y is dangerous is

$$\pi_{(X, Y)} = \min\ (\mu_{FAST}(X), \mu_{DANGEROUS}(Y)).$$

The propositions "X is fast and Y is dangerous" and "If X is fast then Y is dangerous" are translated into the possibility distribution functions, presented in Tables 6.14 and 6.15, respectively.

3. *Quantification rules.* Quantified propositions may have a form "Q Xs are

Table 6.15 "If X is fast then Y is dangerous"

						U_2				
		0	10	20	30	40	50	60	70	80
	0	1	1	1	1	1	1	1	1	1
	10	0.99	1	1	1	1	1	1	1	1
	20	0.98	1	1	1	1	1	1	1	1
	30	0.95	1	1	1	1	1	1	1	1
U_1	40	0.9	0.95	1	1	1	1	1	1	1
	50	0.6	0.65	0.7	0.75	0.8	0.9	1	1	1
	60	0.2	0.25	0.3	0.35	0.4	0.5	0.9	1	1
	70	0.1	0.15	0.2	0.25	0.3	0.4	0.8	1	1
	80	0	0.05	0.1	0.15	0.2	0.3	0.7	1	1

As", where Q is a fuzzy quantifier. Q is interpreted as a fuzzy restriction on the *fuzzy count* Count (A) of the fuzzy set A. In the case of the finite universe U, Count (A) is equal to

$$\sum_{u \in U} \mu_A(u),$$

rounded to the nearest integer, if necessary.

The quantified proposition "Q Xs are As" is translated into

$$\Pi_{\text{Count}(A)} = Q,$$

where $\Pi_{\text{count }(A)}$ represents the possibility distribution of Count (A).

For example, the proposition "Many salaries are high", where the quantifier "many" is defined by the following fuzzy set:

$$\text{MANY} = \{(0, 0), (1, 0.1), (2, 0.2), (3, 0.4), (4, 0.6), (5, 0.7), (6, 0.9), (7, 1)\}$$

is translated into the following possibility distribution:

$$\Pi\text{Count}(\text{HIGH}) = \text{MANY}.$$

For a specific fuzzy set HIGH = $\{(\$10K, 0.1), (\$30K, 0.5), (\$50K, 0.8), (\$70K, 1)\}$, Count (HIGH) = $0.1 + 0.5 + 0.8 + 1 = 2.4$, after rounding Count (HIGH) = 2. Thus,

$$\pi_{\text{Count}(\text{HIGH})}(2) = \mu_{\text{MANY}}(2) = 0.2.$$

The preceding way of translating quantified propositions may be extended to conditional quantified propositions, like "Q As are Bs", where Q is a fuzzy quantifier and A and B are fuzzy sets over the universe U. This proposition is translated into

$$\Pi_{\text{Prop}(B|A)} = Q,$$

where Prop $(B|A)$ or *proportion of* B *in* A, is defined by

$$\text{Prop } (B|A) = \frac{\text{Count } (A \cap B)}{\text{Count } (A)} = \frac{\displaystyle\sum_{u \in U} \min(\mu_A(u), \mu_B(u))}{\displaystyle\sum_{u \in U} \mu_A(u)}.$$

4. *Qualification rules.* Qualified propositions may be qualified by linguistic truth, probability, and possibility.

a. *Truth qualification.* The concept of truth in PRUF enables us to assess the consistency or compatibility of two propositions. Thus, it is not related to the real world. Let p be a proposition "X is A" and q be a proposition "X is B". The *consistency of* p *with* q, denoted Cons (p, q), is given by the possibility that X is A given that X is B, or conversely. Thus,

$$\text{Cons}(p, q) = \sup_{u \in U}(\min(\mu_A(u), \mu_B(u)).$$

For example, if *p* is a proposition "a car is fast" and *q* is a proposition "a car is dangerous", then

$$\text{Cons}(p, q) = \max(\min(0, 0), \min(0.01, 0.05), \min(0.02, 0.1),$$
$$\min(0.02, 0.1), \min(0.05, 0.15), \min(0.1, 0.2), \min(0.4, 0.3),$$
$$\min(0.8, 0.7), \min(0.9, 1), \min(1, 1)) = 1.$$

Suppose *q* is a *reference proposition*, denoted *r*. *The truth of* p *relative to* r is the consistency of *p* with *r*. The truth of *p* relative to *r* may also be defined through a fuzzy subset of [0, 1], called *compatibility of* p *relative to* r, denoted Comp (*p/r*), whose membership function is defined as follows:

$$\text{Comp}(p/r) = \mu_A(B) = \mu_c,$$

where $\mu_A(B)$ is further defined by

$$\mu_r(v) = \sup_{u \in [0, 1]}\mu_B(u),$$

where

$$\mu_A(u) = v,$$

for $v \in [0, 1]$. Thus, if μ_A is one-to-one, then

Table 6.16 Fuzzy Set TRUTH of "car is dangerous" Relative to "car is fast"

u	$\mu_{\text{DANGEROUS}}(\text{FAST})$
0	0
0.05	0.01
0.1	0.02
0.15	0.05
0.2	1
0.3	0.4
0.7	0.8
1	1

$$\mu_\tau(v) = \mu_B(\mu_A^{-1}(v)),$$

provided $\mu_A^{-1}(v) \neq \emptyset$.

The compatibility Comp (p/r) of p relative to r is also called the *truth of* p *relative to* r.

Table 6.16 represents the fuzzy subset TRUTH of "car is dangerous" relative to "car is fast". In the example,

$$\mu_{TRUTH}(1) = 1$$

because

$$\mu_{DANGEROUS}(70) = \mu_{DANGEROUS}(80) = 1,$$

and

$$\mu_{TRUTH}(1) = \sup (\mu_{FAST}(70), \mu_{FAST}(80)) = \sup (0.9, 1) = 1.$$

A truth-qualified proposition has a typical form "X is A is τ", where τ is a fuzzy subset TRUTH of the proposition "X is A" relative to a reference proposition "X is R".

The proposition "X is A is τ" is semantically equivalent to the proposition "X is R", that is,

$$\mu_\tau = \mu_A(R).$$

The proposition "X is R" is translated into the possibility distribution described by

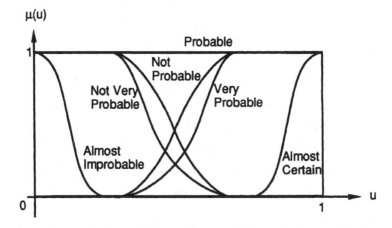

Figure 6.8 Terms of *Probability*

$$\mu_R(u) = \mu_\tau(\mu_A(u))$$

for all $u \in U$. In particular, if τ is the truth value defined by $\mu_\tau(u) = u$ for any $u \in [0, 1]$ (see Figure 6.7), then

$$\mu_R(u) = \mu_A(u)$$

for all $u \in U$.

  b. *Probability qualification.* A typical proposition qualified by fuzzy probability is "X is A is λ", where λ is a term of the linguistic variable "Probability". Such terms are "probable", "not probable", "very probable", or "likely", "not likely", "very likely", and so on. An example of terms of the linguistic variable "Probability" is presented in Figure 6.8.

  There is a close connection between fuzzy probabilities and fuzzy quantifiers. A fuzzy quantifier may be viewed as a fuzzy characterization of the fuzzy count or fuzzy proportion. Thus, a proposition "Q As are Bs', where Q is a fuzzy quantifier, may be interpreted that Q is equal to the fuzzy probability, the conditional probability of A given B.

  Therefore, translations of propositions qualified by fuzzy probabilities may be done like those for quantified propositions.

  c. *Possibility qualification.* Propositions qualified by fuzzy possibilities have the form "X is A is ω", where ω is a term of the linguistic variable "Possibility", such as "possible", "not possible", "very possible", and so on. Again, translations of propositions qualified by fuzzy possibilities may be done by analogy with quantified propositions.

6.1.6.3 Semantic Entailment

Let p and q be propositions, let Π_p and Π_q be possibility distributions induced by p and q, and let π_p and π_q be possibility distribution functions of Π_p and Π_q, respectively. Let U be a universe of discourse. Then, propositions p and q are said to be *semantically equivalent*, denoted $p \leftrightarrow q$, if and only if

$$\pi_p(u) = \pi_q(u)$$

for all $u \in U$.

  For example, the following two propositions are semantically equivalent:

  "Econobox is a fast car is true"

and

  "Econobox is not a fast car is false"

where

$$\mu_{NOT\_FAST}(u) = 1 - \mu_{FAST}(u)$$

for all $u \in U$ and

$$\mu_{FALSE}(v) = \mu_{TRUE}(1 - v)$$

for all $v \in [0, 1]$. Moreover, the preceding two propositions are semantically equivalent for arbitrary fuzzy sets FAST and TRUE. In other words, they are semantically equivalent for all possible denotations for labels of corresponding fuzzy sets. In such a case, the semantic equivalence between p and y is called *strong*.

Two propositions,

"Econobox is a fast car is very true"

and

"Econobox is a very fast car is true"

are semantically equivalent when TRUE is defined by $\mu_{TRUE}(v) = v$ for all $v \in [0, 1]$ but not true for any fuzzy set TRUE on $[0, 1]$. Therefore, the semantic equivalence between the two propositions is not strong.

Using the concept of semantic equivalence, the following proportions may be stated: Let m be a linguistic modifier. If the proposition p has the form "X is A", then

$$m(X \text{ is } A) \leftrightarrow X \text{ is } mA.$$

In particular,

not $(X$ is A and Y is $B) \leftrightarrow X$ is not A or Y is not B,

very $(X$ is A and Y is $B) \leftrightarrow X$ is very A and Y is very B.

For quantified propositions,

$$m(Q \text{ } X\text{'s are } A\text{'s}) \leftrightarrow (mQ) \text{ } X\text{'s are } A\text{'s}.$$

In particular,

not $(Q$ X's are A's$) \leftrightarrow ($not $Q)$ X's are A's.

For qualified propositions,

$$m(X \text{ is } A \text{ is } \tau) \leftrightarrow X \text{ is } A \text{ is } m\tau.$$

Proposition p *semantically entails* proposition q, denoted $p \rightarrow q$, if and only if

$$\pi_p(u) \leq \pi_q(u)$$

for all $u \in U$. This condition may be expressed as

$$A \subseteq B,$$

where A and B are fuzzy subsets of U, $\Pi_p = A$, and $\Pi_q = B$.

For example, the proposition

"Econobox is a very fast car"

semantically entails the proposition

"Econobox is a fast car".

The first proposition semantically entails the second for any possible fuzzy set FAST, so that the semantic entailment is *strong*. However, the proposition

"Econobox is a not fast car"

may semantically entail the proposition

"Econobox is a slow car",

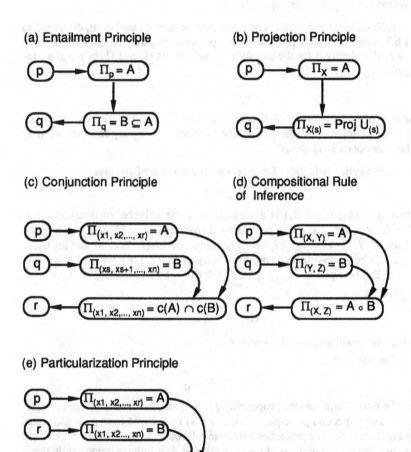

(a) Entailment Principle

$p \longrightarrow \Pi_p = A$

$q \longleftarrow \Pi_q = B \subseteq A$

(b) Projection Principle

$p \longrightarrow \Pi_X = A$

$q \longleftarrow \Pi_{X(s)} = \text{Proj } U_{(s)}$

(c) Conjunction Principle

$p \longrightarrow \Pi_{(x1, x2,..., xr)} = A$

$q \longrightarrow \Pi_{(xs, xs+1,..., xn)} = B$

$r \longleftarrow \Pi_{(x1, x2,..., xn)} = c(A) \cap c(B)$

(d) Compositional Rule of Inference

$p \longrightarrow \Pi_{(X, Y)} = A$

$q \longrightarrow \Pi_{(Y, Z)} = B$

$r \longleftarrow \Pi_{(X, Z)} = A \circ B$

(e) Particularization Principle

$p \longrightarrow \Pi_{(x1, x2,..., xr)} = A$

$r \longrightarrow \Pi_{(x1, x2..., xn)} = B$

$q \longleftarrow \Pi_{(x1, x2,..., xn)} = c(A) \cap B$

Figure 6.9 Rules of Inference in Fuzzy Logic

depending on the definitions of fuzzy sets FAST and SLOW, but the semantic entailment is no longer strong.

6.1.6.4 Rules of Inference

The main rules of inference in fuzzy logic are entailment principle, projection principle, and conjunction principle, with the particularization principle as a special case of the conjunction one. The projection and conjunction principles combined together form the compositional rule of inference, which includes the classical modus ponens as a special case.

1. *Entailment principle.* The concept of semantic entailment, discussed in 6.1.6.3, says that from a proposition p a proposition q may be inferred, denoted $p \to q$, if and only if for the possibility distributions Π_p and Π_q of p and q, respectively,

$$\pi_p(u) \le \pi_q(u)$$

for all u from the universe of discourse U. The idea is illustrated by Figure 6.9 (a). For example, from the proposition "Econobox is a very fast car" we may infer "Econobox is a fast car".

2. *Projection principle.* Let p be a proposition translated into

$$\Pi_X = A,$$

where $X = (X_1, X_2,..., X_n)$ is a variable over the universe of discourse $U = U_1 \times U_2 \times \cdots \times U_n$; Π_X is a possibility distribution over U; and A is a fuzzy subset of U. Let $(i_1, i_2,..., i_k)$ be a subsequence of $(1, 2,..., n)$ and let $(i_{k+1}, i_{k+2},..., i_n)$ be a subsequence complementary to $(i_1, i_2,..., i_k)$, where $1 \le k \le n - 1$. Let $X_{(s)}$ be equal to $(X_{i_1}, X_{i_2},..., X_{i_k})$; let $U_{(s)}$ be equal to $U_{i_1} \times U_{i_2} \times \cdots \times U_{i_k}$; and let

$$\Pi_{X_{(s)}} = \text{Proj } U_{(s)}$$

denote the possibility distribution of $X_{(s)}$.

The equation

$$\Pi_{X_{(s)}} = \text{Proj } U_{(s)}$$

may be retranslated into the proposition q (i.e., q may be translated into the equation). The projection principle says that q may be inferred from p, as illustrated by Figure 6.9(b). The projection principle follows from the entailment principle because, from the definition of the projection of A, it follows immediately that

$$\pi_{X_{(s)}}(u_{(s)}) \ge \pi_X(u)$$

for all $u \in U$.

3. *Conjunction principle.* Let p and q be propositions translated into

$$\Pi_X = A \text{ and } \Pi_Y = B,$$

where $X = (X_1, X_2,..., X_s, X_{s+1},..., X_r)$; $Y = (X_s, X_{s+1},..., X_r, X_{r+1},..., X_n)$; X_1, $X_2,..., X_n$ are variables over the universe of discourse $U = U_1 \times U_2 \times \cdots \times U_n$; $1 \le s \le r \le n$; Π_X is a possibility distribution over $U_1 \times U_2 \times \cdots \times U_r$; A is a fuzzy subset of $U_1 \times U_2 \times \cdots \times U_r$; Π_Y is a possibility distribution over $U_s \times U_{s+1} \times \cdots \times U_n$; and B is a fuzzy subset of $U_s \times U_{s+1} \times \cdots \times U_n$. The conjunction principle states that from propositions p and q a proposition r may be inferred, where r is retranslated from the equation

$$\Pi_{(X_1, X_2,..., X_n)} = c(A) \cap c(B),$$

where $c(A)$ and $c(B)$ are cylindrical extensions of A and B. Again, the conjunction principle follows directly from the entailment principle.

4. *Particularization principle.* The notation is taken from (5). The particularization principle is a special case of the conjunction principle, in which $s = 1$, i.e. $Y = (X_1, X_2,..., X_n)$. Then $c(B) = B$, and the particularization principle asserts that from propositions p and q a proposition r may be inferred, where r is retranslated from the equation

$$\Pi_{(X_1, X_2,..., X_n)} = c(A) \cap B.$$

The particularization principle is illustrated by Figure 6.9(d).

5. *Compositional rule of inference.* The compositional rule of inference is a combination of successive applications of the conjunction principle and the projection principle. As follows from the definition, the composition of fuzzy sets A and B is the projection of the join of cylindrical extensions of A and B. Thus, if p and q are propositions translated into

$$\Pi_{(X, Y)} = A \text{ and } \Pi_{(Y, Z)} = B,$$

where (X, Y) denotes r-tuple of variables $X_1, X_2,..., X_s, X_{s+1},..., X_r$; (Y, Z) denotes $(n - s + 1)$-tuple of variables $X_s, X_{s+1},..., X_r, X_{r+1},..., X_n$; the only common components in (X, Y) and (Y, Z) are $X_s, X_{s+1},..., X_r$; (X, Y) and (Y, Z) are variables over the universes of discourse $U_1 \times U_2 \times \cdots \times U_r$, and $U_s \times U_{s+1} \times \cdots \times U_n$, respectively; $1 \le s \le r \le n$; $\Pi_{(X, Y)}$ is a possibility distribution over $U_1 \times U_2 \times \cdots \times U_r$; $\Pi_{(Y, Z)}$ is a possibility distribution over $U_s \times U_{s+1} \times \cdots \times U_n$; A is a fuzzy subset of $U_1 \times U_2 \times \cdots \times U_r$; and B is a fuzzy subset of $U_s \times U_{s+1} \times \cdots \times U_n$. The compositional rule of inference says that from propositions p and q a proposition r may be inferred where r is retranslated from the equation

$$\Pi_{(X, Z)} = A \circ B.$$

The compositional rule of inference is presented in Figure 6.9(d). Note that the compositional rule of inference includes the classical modus ponens as a special case. To show that, suppose that p is a proposition "X is A" and q is a

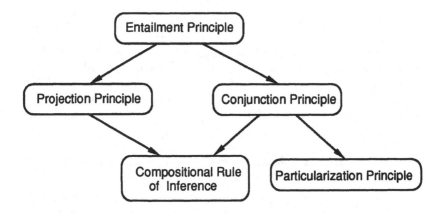

Figure 6.10 Hierarchy of Inference Rules in Fuzzy Logic

proposition "If X is B then Y is C". From the compositional rule of inference and translational rules for composition, a proposition r may be inferred where r is retranslated from

$$\Pi_Y = A \circ (c(B) \oplus c(C)),$$

where $\oplus$ is the bounded sum from Exercise 5. That rule for $A = B$ and crisp (i.e., non fuzzy) sets A, B, C is reduced to the classical modus ponens.

As L. A. Zadeh observed (1981), in probability theory the probability distribution of Y may be induced from the probability distribution of X and the conditional probability of Y given X. In the case of possibility theory, the possibility distribution of X may be inferred from the possibility distribution of Y and the conditional possibility distribution of Y given X.

The generalized modus ponens, expressed previously, has two important properties. First, in the propositions "X is A" and "If X is B then Y is C", all A, B, and C are fuzzy sets. Secondly, A and B need not be identical.

The hierarchy of the preceding inference rules is presented in Figure 6.10.

6.2 Incidence Calculus

Incidence calculus, a tool used to deal with uncertainty, was invented by A. Bundy (1985, 1986). The main idea is to assign sets, called incidences, to propositions instead of probabilities.

Mechanisms to make inferences under uncertainty, such as Bayes' rule, Dempster-Shafer theory, and certainty factors, involve assigning numbers to facts and production rules of expert systems and to functions associated with rules of inference. New numbers, assigned by such functions, are assigned to facts and production rules. Such mechanisms are called *purely numeric* (Bundy, 1985).

Logical connectives of uncertainty mechanisms in which uncertainty measures associated with compound propositions may be computed from the uncertainty measures of their simple propositions are called *truth functional* with respect to the uncertainty measures. Obviously, logical connectives are truth functional with respect to true and false truth values. As A. Bundy observed (1985), logical connectives are not truth functional with respect to probabilities. If probabilities are associated with propositions, then

$$p(f) = 0,$$

$$p(t) = 1,$$

$$p(\neg A) = 1 - p(A),$$

$$p(A \vee B) = p(A) + p(B) - p(A \wedge B),$$

$$p(A \wedge B) = p(A) \cdot p(B) \text{ provided } A \text{ and } B \text{ are independent,}$$

where A and B are propositions, f is a fallacy, and t is a tautology.

The troublemaker is the formula for probability of the conjunction of two propositions A and B. The unconditional alternative for the preceding equation is

$$p(A \wedge B) = p(A) \cdot p(B) + c(A, B) \cdot \sqrt{p(A) \cdot p(\neg A) \cdot p(B) \cdot p(\neg B)} \, ,$$

where $c(A, B)$ is the correlation between two propositions, A and B. Correlation $c(A, B)$ varies between -1 and 1. Correlation $c(A, B) = 0$ means that A and B are independent. $c(A, A) = 1$, and $c(A, \neg A) = -1$. Thus a *correlational probability calculus* is a new candidate for the uncertainty mechanism. The question is whether its logical connectives are truth functional. It is an important question, since if the answer is affirmative, it would be sufficient to establish formulas for $c(\neg A, B)$, $c(A \vee B, C)$, $c(A \wedge B, C)$ from $p(A)$, $p(B)$, $p(C)$, $c(A, B)$, $c(A, C)$, and $c(B, C)$. Unfortunately, an example showing that it does not have to work is given in Subsection 6.2.2.

6.2.1 Incidences

The concept of an incidence is that of an event from classical probability theory (see Section 5.1). A. Bundy selects incidence to denote events that are not propositions. An incidence is a set, a subset of the sample space U.

The set U is a possible world, an interpretation, or a universe of discourse. For a proposition A, the *incidence* i(A) is the subset of U containing all elements of U for which A is true. Incidence calculus follows from the following axioms:

$$i(f) = \emptyset,$$

$$i(t) = U,$$

$$i(\neg A) = -i(A),$$

$$i(A \vee B) = i(A) \cup i(B),$$

$$i(A \wedge B) = i(A) \cap i(B),$$

$$i((\forall x)(A(x)) \subseteq i(A(s)) \subseteq i((\exists x)(A(x)),$$

where $-i(A)$ denotes $U - i(A)$ and s is a term containing no variables not bound in $A(x)$. Unlike equations for probabilities, there is no condition saying that A and B must be independent in the preceding equations. Therefore, all logical connectives are truth functional with respect to incidences.

Let J be a member of U, and let $\zeta(J)$ be the probability of J happening. For a subset I of U, the *weighted probability wp* (I) of I is

$$wp\ (I) = \sum_{J \in I} \zeta(J)$$

For the universe U, $wp(U) = 1$. For a proposition A, let $p(A)$ be the probability of A happening. It is defined as follows:

$$p(A) = wp(i(A)).$$

Let A and B be propositions, and let $p(A|B)$ be the conditional probability of A given B. Then

$$p(A|B) = \frac{wp(i(A) \cap i(B))}{wp(i(B))}.$$

Moreover,

$$p(f) = wp\ (i(f)) = wp\ (\emptyset) = 0,$$

(a)   (b)

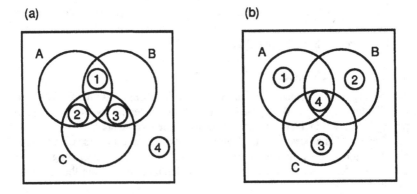

Figure 6.11 Two Cases of Incidences of Propositions A, B, and C

$$p(t) = wp \ (i(t)) = wp \ (U) = 1,$$

$$p(\neg A) = wp \ (i(\neg A)) = wp \ (U - i(A)) = wp \ (U) - wp \ (i(A)) = 1 - wp \ (i(A)) = 1 - p(A),$$

$$p(A \lor B) = wp \ (i(A \lor B)) = wp \ (i(A)) + wp \ (i(B)) - wp \ (i(A \land B)) = p(A) + p(B) - p(A \land B).$$

If A and B are independent, then

$$p(A \land B) = wp \ (i(A \land B)) = wp \ (i(A)) \cdot wp \ (i(B)) = p(A) \cdot p(B).$$

For conditional probability,

$$p(A|B) = \frac{wp \ (i(A) \cap i(B))}{wp \ (i(B))} = \frac{wp \ (i(A \land B))}{wp \ (i(B))} = \frac{p(A \cap B)}{p(B)} \ .$$

6.2.2 Probability Calculus Enhanced by a Correlation

A. Bundy gave an example showing that logical connectives of probability calculus enhanced by a correlation are not truth functional (Bundy, 1985). Here, a simpler example is given for the same purpose. Although a more general result was shown in Dubois and Prade (1988a), namely, that logical connectives of a calculus of uncertainty based on precise reasoning and uncertain information are not truth functional, this example serves also as an illustration of the definition of incidences. In order to show the preceding conjecture, it is enough to give an example of two different cases for propositions A, B, and C such that in both cases $p(A), p(B), p(C), c(A, B), c(A, C)$, and $c(B, C)$ are identical but $c(A \land B, C)$ are different.

Let $U = \{1, 2, 3, 4\}$ and let $\zeta(1) = \zeta(2) = \zeta(3) = \zeta(4) = 0.25$. The incidences of propositions $A, B, C, A \land B$ etc., are represented in Figure 6.11. For example, $i(A) = \{1, 2\}$ in case (a). Therefore, in both cases

$$p(A) = p(B) = p(C) = 0.5,$$

$$p(\neg A) = p(\neg B) = p(\neg C) = 0.5,$$

$$p(A \land B) = p(A \land C) = p(B \land C) = 0.25,$$

$$p(\neg(A \land B)) = p(\neg(A \land C)) = p(\neg(B \land C)) = 0.75,$$

$$c(A, B) = \frac{p(A \land B) - p(A) \cdot p(B)}{\sqrt{p(A) \cdot p(\neg A) \cdot p(B) \cdot p(\neg B)}} = \frac{0.25 - 0.5 \cdot 0.5}{\sqrt{0.5 \cdot 0.5 \cdot 0.5 \cdot 0.5}}$$

$$= 0,$$

and

$$c(A, C) = c(B, C) = 0.$$

In general,

$$c(A \wedge B, C) = \frac{p(A \wedge B \wedge C) - p(A \wedge B) \cdot p(C)}{\sqrt{p(A \wedge B) \cdot p(\neg(A \wedge B)) \cdot p(C) \cdot p(\neg C)}}$$

In case (a),

$$c(A \wedge B, C) = \frac{0 - 0.25 \cdot 0.5}{\sqrt{0.25 \cdot 0.75 \cdot 0.5 \cdot 0.5}} = -0.577,$$

in case (b),

$$c(A \wedge B, C) = \frac{0.25 - 0.25 \cdot 0.5}{\sqrt{0.25 \cdot 0.75 \cdot 0.5 \cdot 0.5}} = 0.577.$$

As Bundy observes, for a pair of propositions X and Y, neither conditional probability nor any other function $d(X, Y)$ would enable $p(X \wedge Y)$ to be computed from $p(X)$, $p(Y)$, and $d(X, Y)$, as the same example shows.

6.2.3 Inference

In general, in a calculus for uncertainty reasoning, if in an inference step the value of uncertainty measure for a proposition A and the implication $A \rightarrow B$ determine the exact value of uncertainty measure of B, then the calculus is *proof functional*. Probability calculus is not proof functional because even if $p(A)$ and $p(B)$ are known, additional knowledge about $c(A, B)$ is necessary for computing $p(A \wedge B)$. In incidence calculus, the measure of uncertainty of a proposition is its incidence. For an implication $A \rightarrow B$, when $i(A)$ is given and modus ponens is used, only $i(A) \subseteq i(B)$ may be inferred, so incidence calculus is not proof functional either.

Each inference for B, using different implications with B as a conclusion, provides a new lower bound for $i(B)$. The greatest lower bound is the union of all these lower bounds. As A. Bundy observes, the lower bounds given by incidence calculus are much tighter than those provided by a purely numeric probability calculus because logical connectives are truth functional with respect to incidences but not to probabilities (Bundy, 1985).

In predicate calculus, nothing may be inferred about $A \wedge B$ from a true proposition A and a proposition B for which the truth value is unknown. In incidence calculus, if $i(A)$ is given, a new upper bound for $i(A \wedge B)$ may be evaluated as $i(A)$.

6.3 Rough Set Theory

A tool to deal with uncertainty, called rough set theory, was introduced by Z. Pawlak (1982). This theory is especially well suited to deal with inconsistencies in the process of knowledge acquisition. In the presented approach, inconsistencies are not corrected. The key issue is to compute lower and upper approximations of partitions, the fundamental concepts of rough set theory. On the basis of lower and upper approximations, two different sets of rules are computed: certain and possible. Certain rules are categorical and may be further employed using classical logic. Possible rules are supported by existing data, although conflicting data may exist as well. Possible rules may be processed further using either classical logic or any theory to deal with uncertainty.

Note that the presented approach may be combined with any other approach to uncertainty. An advantage of the method is that certain and possible rules are processed separately (i.e., two parallel inference engines may be used). Another advantage of rough set theory is that it does not need any preliminary or additional information about data (like prior probability in probability theory, basic probability number in Dempster-Shafer theory, and grade of membership in fuzzy set theory).

6.3.1 Rough Sets

Let U be a nonempty set, called the *universe*, and let R be an equivalence relation on U, called an *indiscernibility relation*. An ordered pair $A = (U, R)$ is called an *approximation space*. For any element x of U, the equivalence class of R containing x is denoted $[x]_R$. Equivalence classes of R are called *elementary sets in* A. We assume that the empty set is also elementary.

Any finite union of elementary sets in A is called a *definable set in* A.

Let X be a subset of U. We wish to define X in terms of definable sets in A. Thus, we need two more concepts.

A *lower approximation of* X *in* A, denoted $\underline{R}X$, is the set

$$\{x \in U \mid [x]_R \subseteq X \}.$$

An *upper approximation of* X *in* A, denoted $\bar{R}X$, is the set

$$\{x \in U \mid [x]_R \cap X \neq \emptyset \}.$$

The lower approximation of X in A is the greatest definable set in A, contained in X. The upper approximation of X in A is the least definable set in A containing X. Time complexity of algorithms for computing lower and upper approximations of any set X is $O(m^2)$, where m is the cardinality of set U of entities.

For example, $U = \{x_1, x_2, x_3, x_4, x_5, x_6, x_7, x_8\}$, and R is an equivalence relation that induces the following partition:

$$R* = \{\{x_1\}, \{x_2, x_3\}, \{x_4, x_5\}, \{x_6, x_7\}, \{x_8\}\},$$

that is, the equivalence classes of R are $\{x_1\}$, $\{x_2, x_3\}$, $\{x_4, x_5\}$, $\{x_6, x_7\}$, and $\{x_8\}$. Let X be equal to $\{x_1, x_2, x_3, x_5, x_7\}$. Then the lower approximation of X in $A = (U, R)$ is the set

$$\underline{R}X = \{x_1, x_2, x_3\},$$

and the upper approximation of X in A is the set

$$\bar{R}X = \{x_1, x_2, x_3, x_4, x_5, x_6, x_7\},$$

(a) Lower Approximation of X in A

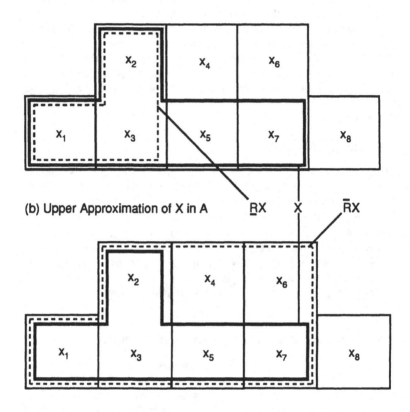

Figure 6.12 Sets X, $\underline{R}X$, and $\bar{R}X$

(see Figure 6.12).

Let X and Y be subsets of U. Lower and upper approximations of X and Y in A have the following properties

$$\underline{R}X \subseteq X \subseteq \bar{R}X,$$

$$\underline{R}U = U = \bar{R}U,$$

$$\underline{R}\emptyset = \emptyset = \bar{R}\emptyset,$$

$$\underline{R}(X \cup Y) \supseteq \underline{R}X \cup \underline{R}Y,$$

$$\bar{R}(X \cup Y) = \bar{R}X \cup \bar{R}Y,$$

$$\underline{R}(X \cap Y) = \underline{R}X \cap \underline{R}Y,$$

$$\overline{R}(X \cap Y) \subseteq \overline{R}X \cap \overline{R}Y,$$

$$\underline{R}(X - Y) \subseteq \underline{R}X - \underline{R}Y,$$

$$\overline{R}(X - Y) \supseteq \overline{R}X - \overline{R}Y,$$

$$\underline{R}(-X) = -\overline{R}X,$$

$$\overline{R}(-X) = -\underline{R}X,$$

$$\underline{R}X \cup \overline{R}(-X) = X,$$

$$\underline{R}(\underline{R}X) = \overline{R}(\underline{R}X) = \underline{R}X,$$

$$\overline{R}(\overline{R}X) = \underline{R}(\overline{R}X) = \overline{R}X,$$

where $-X$ denotes the complement $U - X$ of X.

A *rough set in* A is the family of all subsets of U having the same lower and upper approximations in A. In the example from Figure 6.12, the rough set is equal to

$$\{\{x_1, x_2, x_3, x_4, x_6\}, \{x_1, x_2, x_3, x_4, x_7\},$$
$$\{x_1, x_2, x_3, x_5, x_6\}, \{x_1, x_2, x_3, x_5, x_7\}\}.$$

Let x be in U. We say that x is *certainly in* X if and only if $x \in \underline{R}X$, and that x is *possibly in* X if and only if $x \in \overline{R}X$. Our terminology originates from the fact that we want to decide if x is in X on the basis of a definable set in A rather than on the basis of X. This means that we deal with $\underline{R}X$ and $\overline{R}X$ instead of X. Since $\underline{R}X \subseteq X \subseteq \overline{R}X$, if x is in $\underline{R}X$ it is certainly in X. On the other hand, if x is in $\overline{R}X$, it is possibly in X. In the example from Figure 6.12, x_3 is certainly in X, and therefore possibly in X, x_4 is possibly but not certainly in X, and x_8 is not possibly in X.

6.3.2 Rough Definability of a Set

A rough-set classification of a set of entities has been presented in Pawlak (1984). The main idea is to classify any such set by properties of its lower and upper approximations. The notation of Chapter 3 pertaining to decision tables is used in the sequel. Let Q be a set of all attributes and decisions. Let U be a set of all entities. For any nonempty subset P of Q, an ordered pair $(U, \overset{\frown}{P})$ is an approximation space A. For the sake of convenience, for any $X \subseteq U$, the lower approximation of X in A and the upper approximation of X in A is called P-*lower approximation of* X and P-*upper approximation of* X, and is denoted $\underline{P}X$ and $\overline{P}X$, respectively.

An example of a decision table is presented in Table 6.17. Let $Q = \{a, b, c, d\}$ be the set of all attributes and decisions. Let P be the set of all attributes, i.e., $P = \{a, b, c\}$. The partition generated by P, P^*, is equal to

$$\{\{x_1\}, \{x_2, x_3\}, \{x_4, x_5\}, \{x_6, x_7\}, \{x_8\}\}.$$

Table 6.17 A Decision Table

| | Attributes | | | Decision |
	a	b	c	d
x_1	0	3	0	0
x_2	0	4	1	0
x_3	0	4	1	0
x_4	1	4	1	1
x_5	1	4	1	0
x_6	2	4	1	1
x_7	2	4	1	0
x_8	2	5	2	1

Let X be a set $\{x_1, x_2, x_3, x_5, x_7\}$. Then

$$\underline{P}X = \{x_1, x_2, x_3\}$$

and

$$\bar{P}X = \{x_1, x_2, x_3, x_4, x_5, x_6, x_7\}.$$

Any union of equivalence classes of $\underset{P}{\sim}$ is called a P-*definable set*. A set $X \subseteq U$ which is not P-definable is called P-*undefinable*.

The set X is called *roughly* P-*definable* if and only if $\underline{P}X \neq \emptyset$ and $\bar{P}X \neq U$.

The set X is called *internally* P-*undefinable* if and only if $\underline{P}X = \emptyset$ and $\bar{P}X \neq U$.

The set X is called *externally* P-*undefinable* if and only if $\underline{P}X \neq \emptyset$ and $\bar{P}X = U$.

The set X is called *totally* P-*undefinable* if and only if $\underline{P}X = \emptyset$ and $\bar{P}X = U$.

For an internally P-undefinable set X we can not say certainly that any $x \in U$ is a member of X. For an externally P-undefinable set X we can not exclude any element $x \in U$ being possibly a member of X.

Note that a set X may be P-undefinable and roughly P-definable. Yet another possibility is a set X that is P-definable and not roughly P-definable (e.g., the empty set $\emptyset$). If a set X is not roughly P-definable, then it is internally, externally, or totally P-undefinable. In the example from Table 6.17, set $\{x_1, x_2,$

x_3} is P-definable, set X is roughly P-definable, set {x_3, x_5, x_7} is internally P-undefinable, and set {x_1, x_3, x_5, x_7, x_8} is externally P-undefinable. In the example a P-undefinable set does not exist.

6.3.3 Rough Measures of a Set

Measures of uncertainty based on rough set theory have auxiliary value only. In the rough-set approach, the set X is described by its lower and upper approximations.

A *quality of lower approximation of* X *by* P, denoted $\gamma_P(X)$, is equal to

$$\frac{|\underline{P}X|}{|U|} .$$

In Pawlak (1984), $\gamma_P(X)$ is called quality of approximation of X by P.

Thus, the quality of lower approximation of X by P is the ratio of the number of all certainly classified entities by attributes from P as being in X to the number of all entities of the system. It is a kind of relative frequency. Note that $\gamma_P(X)$ is a belief function according to Dempster-Shafer theory (Shafer, 1976) because it satisfies all three conditions of the definition of belief function. There are essential connections between rough set theory and Dempster-Shafer theory. For example, lower and upper approximations of rough set theory exist under the names of inner and outer reductions, respectively, in the monograph by G. Shafer (1976). However, both theories have been developed separately and both have well-established records of publications. The main difference is in the emphasis: Dempster-Shafer theory uses belief function as a main tool, while rough set theory makes use of the rough set, i.e., the family of all sets with common lower and upper approximations.

Let $P^* = \{U_1, U_2, ..., U_k\}$ be a partition of U generated by P. It is not difficult to recognize that the basic probability assignment m that corresponds to γ_P is given by

$$m(U_i) = \frac{|U_i|}{|U|} ,$$

where $i = 1, 2, ..., k$, and $m(Y) = 0$ for all other $Y \subseteq U$. Therefore, focal elements (Shafer, 1976) of γ_P are elementary sets in the approximation space $(U, \stackrel{\frown}{P})$, and the core of γ_P is the set U.

A *quality of upper approximation of* X *by* P, denoted $\overline{\gamma}_P(X)$, is

$$\frac{|\overline{P}X|}{|U|} .$$

The quality of upper approximation of X by P is the ratio of the number of all possibly classified entities by attributes from P as being in X to the number of all entities of the system. Therefore, it is again a kind of relative frequency.

The quality of upper approximation of X by P is a plausibility function from the Dempster-Shafer theory viewpoint (Shafer, 1987), called an upper probability function in Shafer (1976). In the example from Table 6.17,

$$\gamma_P(X) = \frac{3}{8} = 0.375, \text{ and } \overline{\gamma}_P(X) = \frac{7}{8} = 0.875.$$

6.3.4 Partitions

In this section, the concept of rough definability will be expanded to include partitions of the set of all entities. A partition is of great interest because in the process of learning from examples, rules are derived from partitions, generated by single decisions.

It is quite natural to expect that all concepts of rough definability of a set may be transformed into similar ones for partitions. Surprisingly, this is not true. For example, as follows from Grzymala-Busse (1988), externally undefinable partitions do not exist. Instead, they coalesce with totally undefinable partitions.

Let $\mathcal{X} = \{X_1, X_2, ..., X_n\}$ be a partition of U. P-*lower* and P-*upper approximations of $\mathcal{X}$ by* P, denoted $\underline{P}\mathcal{X}$ and $\overline{P}\mathcal{X}$, respectively, are the following sets:

$$\underline{P}\mathcal{X} = \{\underline{P}X_1, \underline{P}X_2, ..., \underline{P}X_n\}$$

and

$$\overline{P}\mathcal{X} = \{\overline{P}X_1, \overline{P}X_2, ..., \overline{P}X_n\}.$$

The partition $\mathcal{X}$ of U is called P-*definable* if and only if $\underline{P}\mathcal{X} = \overline{P}\mathcal{X}$. The partition $\mathcal{X}$ of U that is not P-definable is called P-*undefinable*.

The partition $\mathcal{X}$ of U is called *roughly* P-*definable, weak* if and only if there exists a number $i \in \{1, 2, ..., n\}$ such that $\underline{P}X_i \neq \emptyset$. For a roughly P-definable, weak partition $\mathcal{X}$ of U, there exists a number $j \in \{1, 2, ..., n\}$ such that $\overline{P}X_j \neq U$ (Grzymala-Busse, 1988).

The partition $\mathcal{X}$ of U is called *roughly* P-*definable, strong* if and only if $\underline{P}X_i \neq \emptyset$ for each i from the set $\{1, 2, ..., n\}$. In a roughly P-definable, strong partition $\mathcal{X}$ of U, $\overline{P}X_i \neq U$ for each i from the set $\{1, 2, ..., n\}$ (Grzymala-Busse, 1988).

The partition $\mathcal{X}$ of U is called *internally* P-*undefinable, weak* if and only if $\underline{P}X_i = \emptyset$ for each $i \in \{1, 2, ..., n\}$ and there exists $j \in \{1, 2, ..., n\}$ such that $\overline{P}X_j \neq U$.

The partition $\mathcal{X}$ of U is called *internally* P-*undefinable, strong* if and only if $\underline{P}X_i = \emptyset$ and $\overline{P}X_i \neq U$ for each i from the set $\{1, 2, ..., n\}$.

Finally, the partition $\mathcal{X}$ of U is called *totally* P-*undefinable* if and only if $\underline{P}X_i = \emptyset$ and $\overline{P}X_i = U$ for each $i \in \{1, 2, ..., n\}$. For example, for the decision table presented in Table 6.18, let $P = \{a, b\}$ and $U = \{x_1, x_2, x_3, x_4, x_5, x_6, x_7, x_8\}$. Then

Table 6.18 A Decision Table

	Attributes		Decisions					
	a	b	c	d	e	f	g	h
x_1	0	0	0	0	0	0	0	0
x_2	0	0	0	0	0	1	1	1
x_3	0	0	0	0	0	2	1	1
x_4	0	1	0	0	0	0	0	0
x_5	0	1	0	1	1	1	2	1
x_6	1	1	1	1	1	0	1	0
x_7	1	1	1	2	1	1	2	0
x_8	1	1	1	2	1	2	2	1

$$\underline{P}\{c\}* = \overline{P}\{c\}* = \{\{x_1, x_2, x_3, x_4, x_5\}, \{x_6, x_7, x_8\}\},$$

$$\underline{P}\{d\}* = \{\{x_1, x_2, x_3\}, \emptyset\}, \ \overline{P}\{d\}* = \{\{x_1, x_2, x_3, x_4, x_5\}, \{x_4, x_5, x_6, x_7, x_8\}\},$$

$$\underline{P}\{e\}* = \{\{x_1, x_2, x_3\}, \{x_6, x_7, x_8\}\}, \ \overline{P}\{e\} = \{\{x_1, x_2, x_3, x_4, x_5\},$$
$$\{x_4, x_5, x_6, x_7, x_8\}\},$$

$$\underline{P}\{f\}* = \{\emptyset\}, \overline{P}\{f\}* = \{U, \{x_1, x_2, x_3, x_6, x_7, x_8\}\},$$

$$\underline{P}\{g\}* = \{\emptyset\}, \overline{P}\{g\}* = \{\{x_1, x_2, x_3, x_4, x_5\}, \{x_1, x_2, x_3, x_6, x_7, x_8\},$$
$$\{x_4, x_5, x_6, x_7, x_8\}\},$$

$$\underline{P}\{h\}* = \{\emptyset\}, \overline{P}\{h\}* = \{U\}.$$

Thus, the partition $\{c\}*$ is P-definable, $\{d\}*$ is roughly P-definable, weak, $\{e\}*$ is roughly P-definable, strong, $\{f\}*$ is internally P-undefinable, weak, $\{g\}*$ is internally P-definable, strong, and $\{h\}*$ is totally P-undefinable.

A *quality of lower approximation of* X *by* P is equal to

$$\frac{\sum_{i=1}^{n}|\underline{P}X_i|}{|U|},$$

and a *quality of upper approximation of* X *by* P is equal to

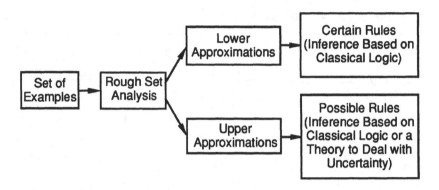

Figure 6.13 The Principle of the Use of Rough Set Theory for Learning from Examples

$$\frac{\sum_{i=1}^{n} |\overline{P}X_i|}{|U|} .$$

In the example from Table 6.18, the quality of lower approximation of $\{d\}^*$ is

$$\frac{3 + 0}{8} = 0.375,$$

and the quality of upper approximation of $\{d\}^*$ is

$$\frac{5 + 5}{8} = 1.25.$$

6.3.5 Certain and Possible Rules

The main idea of use of rough set theory for learning from examples is presented in Figure 6.13. A set of examples is given in the form of a decision table. All results of uncertainty are manifested finally by inconsistent information in the decision table.

Another example of a decision table is given by Table 6.19, which may be interpreted as containing data about patients denoted by numbers 0, 1,..., 9. Table 6.19 includes inconsistencies. For example, patients 3 and 4, described by the same values of attributes, are classified differently by a decision: Patient 3 is free from influenza, while patient 4 has it. Another pair of inconsistencies is introduced by patients 6 and 7. In the example, the partition of patients as being

Table 6.19 A Decision Table

	Attributes				Decision
	Temperature	Dry_cough	Headache	Muscle_pain	Influenza
0	normal	absent	absent	absent	absent
1	normal	absent	present	present	absent
2	subfebrile	absent	present	present	present
3	subfebrile	present	absent	absent	absent
4	subfebrile	present	absent	absent	present
5	high	absent	absent	absent	absent
6	high	present	absent	absent	absent
7	high	present	absent	absent	present
8	high	present	present	present	present
9	high	present	present	present	present

sick with influenza (or, according to the preceding terminology, partition generated by {Influenza}) is

$$\chi = \{\{0, 1, 3, 5, 6\}, \{2, 4, 7, 8, 9\}\}.$$

For the above example,

{Temperature, Dry_cough, Headache, Muscle_pain}*

$=\{\{0\}, \{1\}, \{2\}, \{3, 4\}, \{5\}, \{6, 7\}, \{8, 9\}\}$

$=\{$Temperature, Dry_cough, Headache$\}$*.

Thus, the attribute Muscle_pain is redundant as a symptom. The set {Temperature, Dry_cough, Headache} will be denoted P. The lower approximation $\underline{P}\chi$ is equal to

$$\{\{0, 1, 5\}, \{2, 8, 9\}\}$$

and the upper approximation $\overline{P}\chi$ is equal to

$$\{\{0, 1, 3, 4, 5, 6, 7\}, \{2, 3, 4, 6, 7, 8, 9\}\}.$$

Certain rules follow from $\underline{P}\chi$. Every element of $\underline{P}\chi$ is P-definable; hence, it may be represented by rules using attributes of set P. Furthermore, set {0, 1,

5}, an element of $\underline{P}\chi$, is described by the certain fact (Influenza, absent). Corresponding certain rules are

> (Temperature, normal) $\wedge$ (Dry_cough, absent) $\wedge$ (Headache, absent) $\rightarrow$ (Influenza, absent),
>
> (Temperature, normal) $\wedge$ (Dry_cough, absent) $\wedge$ (Headache, present) $\rightarrow$ (Influenza, absent),
>
> (Temperature, high) $\wedge$ (Dry_cough, absent) $\wedge$ (Headache, absent) $\rightarrow$ (Influenza, absent).

The preceding certain rules, after simplification by dropping conditions, become

> (Temperature, normal) $\rightarrow$ (Influenza, absent),
>
> (Temperature, high) $\wedge$ (Dry_cough, absent) $\rightarrow$ (Influenza, absent).

Another element of $\underline{P}\chi$, set {2, 8, 9}, implies the following certain rule (after simplification):

> (Temperature, not normal) $\wedge$ (Headache, present) $\rightarrow$ (Influenza, present).

Possible rules are generated by $\bar{P}\chi$. Again, every element of $\bar{P}\chi$ is P-definable. On the other hand, set {0, 1, 3, 4, 5, 6, 7}, an element of $\bar{P}\chi$, is described by the possible fact (Influenza, absent). Thus, possible rules, after simplification, are

> (Temperature, normal) $\rightarrow$ (Influenza, absent),
>
> (Headache, absent) $\rightarrow$ (Influenza, absent).

Possible rules that follow from the other element of $\bar{P}\chi$, after simplification, are

> (Temperature, subfebrile) $\rightarrow$ (Influenza, present),
>
> (Dry_cough, present) $\rightarrow$ (Influenza, present).

6.4 Concluding Remarks

Fuzzy set theory is very controversial. On one hand, it is an extremely popular area of research. Successful real-life systems based on fuzzy set theory have been implemented. Moreover, possibility theory, using its language, is most likely the most popular approach to handling uncertainty in expert systems. New applications of fuzzy set theory to expert systems are extensively developed (di Nola *et al.*, 1989), (Dubois and Prade, 1988c). On the other hand, there is a lot of opposition to fuzzy set theory in the AI community, and not only there. In mild form, fuzzy set theory is prohibited from describing uncertainty at all and instead is assumed to be able to deal with ambiguity in describing events (Pearl,

1988). In a stronger form of criticism (Cheeseman, 1986), fundamental rules of fuzzy set theory are seen as false. For example, P. Cheeseman says that the intersection of fuzzy sets A and B over universe U, defined by

$$\mu_{A \cap B}(u) = \min \left(\mu_A(u), \mu_B(u) \right),$$

is wrong, in general. Instead of the preceding formula, Cheeseman suggests that when no sufficient information about A and B is provided, the only conclusion about $\mu_{A \cap B}(u)$ should be

$$0 \leq \mu_{A \cap B}(u) \leq \min \left(\mu_A(u), \mu_B(u) \right),$$

as it is for probabilities.

Some other problems of fuzzy set theory are associated with assigning values for a membership function. Moreover, membership function is context sensitive (e.g., a *small* elephant is *bigger* than a *big* mouse) (Lee *et al.*, 1987).

Among many ways of reasoning based on fuzzy set theory, *possibilistic logic* was first proposed by H. Prade in 1983 (Dubois and Prade, 1988a). In possibilistic logic, a proposition is graded by two numbers, called a *necessity* and a *possibility*. Possibilistic logic, like probability theory or Dempster-Shafer theory, offers an approach to precise reasoning on uncertain information. As was shown in Dubois and Prade (1988a), in any calculus of uncertainty (such as possibilistic logic, probability theory, or Dempster-Shafer theory), its logical connectives are not truth functional, while a fuzzy logic, involved in fuzzy reasoning on precise information, called a *logic of vagueness*, can have truth-functional connectives.

A. Bundy's incidence calculus reflects the opinion that uncertainty should not be represented by a single numerical value. Instead, a set of situations in which the proposition is true is attached to a proposition. According to Mamdani *et al.* (1985), incidence calculus is restricted in applications and may be applied only to some kinds of uncertainty.

Rough set theory and fuzzy set theory are independent, although rough set theory is related to Dempster-Shafer theory. The main advantage of rough set theory is that it does not need any preliminary or additional information about data, yet it is possible to induce certain and possible rules. There exist successful real-life implementations of rough set theory, for example, knowledge-based systems in medicine (Pawlak, 1984) or in industry (Mrozek, 1987; Ziarko and Katzberg, 1989). The learning system LERS, based on rough set theory, has been also implemented (Grzymala-Busse and Sikora, 1988).

Exercises

1. Define in your own way the fuzzy sets corresponding to the following concepts and universal sets:

    a. A high annual salary, $U = \{\$10K, \$20K, ..., \$200K\}$,

    b. A low annual salary, $U = \{\$10K, \$20K, ..., \$200K\}$,

    c. The normal temperature of the human body, U = {98.6°F, 98.8°F,...,
103.0°F},

    d. Fever, U = {98.6°F, 98.8°F,..., 103.0°F},

    e. High fever, U = {98.6°F, 98.8°F,..., 103.0°F},

    f. A thick book, U = {100 pp., 150 pp.,..., 1000 pp.},

    g. A very thick book, U = {100 pp., 150 pp.,..., 1000 pp.}.

2. Determine the complements for all fuzzy sets from Exercise 1.

3. Determine the complements for the following fuzzy sets over the universe set
U = {0, 1,..., 9}:

    a. A = {(0, 0.5), (1, 0.3), (2, 0.1), (3, 0.4), (4, 0.6), (5, 0.9), (6, 1), (7,
0.8), (8, 0.7), (9, 0)}

    b. B = {(0, 0), (1, 0.1), (2, 0.4), (3, 0.7), (4, 0.5), (5, 0.4), (6, 0.2), (7,
0.5), (8, 0.7), (9, 1)}

    c. C = {(0, 1), (1, 0.7), (2, 0.4), (3, 0.2), (4, 0), (5, 0), (6, 0), (7, 0.4), (8,
0.6), (9, 0.8)}.

4. Determine the unions and intersections of the following fuzzy sets:

    a. The fuzzy sets from (a) and (b), Exercise 1,

    b. The fuzzy sets from (c) and (d), Exercise 1,

    c. The fuzzy sets A and B, Exercise 3,

    d. The fuzzy sets A and C, Exercise 3,

    e. The fuzzy sets B and C, Exercise 3.

5. The union of two fuzzy sets A and B may be defined differently than in Subsection 6.1.1., for example, as

    i.   The probabilistic sum (Zadeh, 1965), denoted +, where for all $u \in U$,

$$\mu_{A+B}(u) = \mu_A(u) + \mu_B(u) - \mu_A(u) \cdot \mu_B(u),$$

    ii. The bounded sum, according to L. A. Zadeh, called a bold union in
Giles (1976), denoted $\oplus$, where for all $u \in U$,

$$\mu_{A \oplus B}(u) = \min(1, \mu_A(u) + \mu_B(u)),$$

    iii. Yager's union (Yager, 1980), denoted $\cup$, where for all $u \in U$,

$$\mu_{A \cup B}(u) = \min (1, (\mu_A(u)^p + \mu_B(u)^p)^{1/p}),$$

where $p \geq 1$. For $p \to \infty$, $\mathring{\cup}$ converges to the ordinary union operator from Subsection 6.1.1. For $p = 1$, $\mathring{\cup}$ becomes the bounded sum.

Similarly, the intersection of two fuzzy sets A and B may be defined as

i. The probabilistic product, denoted $\cdot$, where for all $u \in U$,

$$\mu_{A \cdot B}(u) = \mu_A(u) \cdot \mu_B(u),$$

ii. The bounded product (bold intersection), denoted $\odot$, where for all $u \in U$,

$$\mu_{A \odot B}(u) = \max(0, \mu_A(u) + \mu_B(u) - 1),$$

iii. Yager's intersection (Yager, 1980), denoted $\mathring{\cap}$, where for all $u \in U$,

$$\mu_{A \mathring{\cap} B}(u) = 1 - \min(1, ((1 - \mu_A(u))^p + (1 - \mu_B(u))^p)^{1/p}),$$

where $p \geq 1$. For $p \to \infty$, $\mathring{\cap}$ converges to the ordinary union operator from Subsection 6.1.1. For $p = 1$, $\mathring{\cap}$ becomes the bounded product.

    a. Check which of properties (1–9) of 6.1.1. hold for the probabilistic sum and product,

    b. Check which of properties (1–9) of 6.1.1. hold for the bounded sum and product,

    c. Check which of properties (1–9) of 6.1.1. hold for Yager's union and intersection.

6. For the following fuzzy sets: A and B, A and C, and B and C from Exercise 3, determine

    a. Their probabilistic sums and products,

    b. Their bounded sums and products,

    c. Their Yager's union and intersection for $p = 2$.

Definitions of the preceding operations are given in Exercise 5.

7. Define in your own words fuzzy binary relations corresponding to the following concepts and universal sets:

    a. Approximately equal, $U_1 = U_2 = \{0, 1,..., 9\}$,

    b. Likes, $U_1 = \{$Bill, Jan, Su$\}$, $U_2 = \{$Bob, Jan, Jim$\}$,

    c. Faster than, $U_1 = U_2 = \{$bison, fox, lion, llama, opossum, rabbit$\}$.

8. For the following k-tuples $U_{(s)}$ and relations R on U, determine the projections $Proj_{U_{(s)}}(R)$:

a. $U = U_1 \times U_2$, $U_1 = \{a, b, c\}$, $U_2 = \{d, e, f\}$, $R = \{((a, d), 0), ((a, e), 0.2),$
    $((a, f), 0.4), ((b, d), 0.7), ((b, c), 0.5), ((b, f), 0.3), ((c, d), 0.6), ((c, e),$
    $1), ((e, f), 0.8)\}$,

b. U and R as in (a), but $U_{(s)} = U_2$,

c. $U = U_1 \times U_2$, $U_{(s)} = U_1$, $U_1 = U_2 = [1, 4]$, $\mu_R(u, v) = \dfrac{1}{2u} + \dfrac{1}{2v}$, where u
    $\in U_1$ and $v \in U_2$,

d. $U = U_1 \times U_2 \times U_3$, $U_1 = \{a, b\}$, $U_2 = \{b, c\}$, $U_3 = \{c, d\}$, $U_{(s)} = U_1 \times U_2$,
    $R = \{(a, b, c), 0.4), ((a, b, d), 0.6), ((a, c, c), 0.7), ((a, c, d), 0.3), ((b,$
    $b, c), 0.8), (b, b, d), 1), ((b, c, c), 0), ((b, c, d), 0.2)\}$,

e. U and R as in (d), $U_{(s)} = U_2 \times U_3$,

f. U and R as in (d), $U_{(s)} = U_1$,

g. U and R as in (d), $U_{(s)} = U_2$.

9. For the following relations R on $U_{(s)}$, determine the cylindrical extensions on
U, where

a. $U_{(s)} = U_1 = \{a, b, c\}$, $U = U_1 \times U_2$, $U_2 = \{d, e, f\}$, $R = \{(a, 0), (b, 0.4),$
    $(c, 0.8)\}$,

b. $U_{(s)} = U_2 = \{a, b, c\}$, $U = U_1 \times U_2 \times U_3$, $U_1 = U_2 = U_3$, $R = \{(a, 1),$
    $(b, 0.5), (c, 0.1)\}$,

c. $U_{(s)} = U_1 = [0, 1]$, $U = U_1 \times U_2$, $U_2 = [0, 1]$, μ_R is the following S-func-
    tion:

    $$\mu_R(u) = S(u; 0.2, 0.5, 0.8) \text{ for all } u \in U_1,$$

d. $U_{(s)} = U_1 \times U_2$, $U_1 = U_2 = [0, 1]$, $U = U_1 \times U_2 \times U_3$, $U_3 = [0, 1]$, $\mu_R(u,$
    $v) = \dfrac{u + v}{2}$ for $u \in U_1$ and $v \in U_2$.

10. For the following two fuzzy relations R and S, determine the sup-min composition $R \circ S$, where

a. R is defined on $U_1 = \{a, b, c\}$,

    S is defined on $= U_1 \times U_2$, $U_2 = \{d, e\}$,

    $R = \{(a, 0), (b, 0.4), (c, 0.8)\}$,

    $S = \{((a, d), 0.4), ((a, e), 0.2), ((b, d), 0.3), ((b, e), 0.5), ((c, d), 0.9),$
    $((c, e), 0.6)\}$,

b. R is defined on $U_1 \times U_2$,

    S is defined on $U_2 \times U_3$, $U_1 = U_2 = U_3 = \{a, b, c\}$,

    $R = \{((a, a), 0.3), ((a, b), 0.2), ((a, c), 0.8), ((b, a), 0.2), ((b, b), 0.7),$
    $((b, c), 0.7), ((c, a), 0.6), ((c, b), 0.3), ((c, c), 0)\}$,

$S = \{((a, a), 0.2), ((a, b), 0.1), ((a, c), 0.1), ((b, a), 0.4), ((b, b), 0.5),$
$((b, c), 0.8), ((c, a), 0.1), ((c, b), 0.4), ((c, c), 1)\}$,

c. R is defined on $U_1 \times U_2 \times U_3$,

S is defined on $U_2 \times U_3$, $U_1 = U_2 = U_3 = \{0, 1\}$,

$R = \{((0, 0, 0), 0.2), ((0, 0, 1), 0.3), ((0, 1, 0), 0.6), ((0, 1, 1), 0.4),$
$((1, 0, 0), 0.8), ((1, 0, 1), 0.3), ((1, 1, 0), 0.2), ((1, 1, 1), 0.9)\}$,

$S = \{((0, 0), 0.1), ((0, 1), 0.4), ((1, 0), .2), ((1, 1), 0.3)\}$.

d. R is defined on $U_1 \times U_2 \times U_3$,

S is defined on $U_2 \times U_3 \times U_4$, $U_1 = U_2 = U_3 = U_4 = \{0, 1\}$,

R is as in (c),

$S = \{((0, 0, 0), 0.3), ((0, 0, 1), 0.4), ((0, 1, 0), 0.4), ((0, 1, 1), 0), ((1, 0, 0), 0), ((1, 0, 1), 0.8), ((1, 1, 0), 0.1), ((1, 1, 1), 1)\}$,

e. R is defined on $U_1 \times U_2 \times U_3$,

S is defined on $U_3 \times U_4 \times U_5$, $U_1 = U_2 = U_3 = U_4 = U_5 = \{0, 1\}$,

R and S as in (d).

11. Prove that the sup-min composition of relations is associative, that is, $R \circ (S \circ T) = (R \circ S) \circ T$, where R, S, and T are binary fuzzy relations on $U_1 \times U_2$, $U_2 \times U_3$, and $U_3 \times U_4$, respectively.

12. Prove that the sup-min composition of relations is distributive over the union, that is, $R \circ (S \cup T) = (R \circ S) \cup (R \circ T)$, where R, S, and T are binary fuzzy relations on $U_1 \times U_2$, $U_2 \times U_3$, and $U_3 \times U_4$, respectively.

13. Prove that the sup-min composition of relations is weakly distributive over the intersection, that is, $R \circ (S \cap T) \subseteq (R \circ S) \cap (R \circ T)$, where R, S, and T are binary fuzzy relations on $U_1 \times U_2$, $U_2 \times U_3$, and $U_3 \times U_4$, respectively. Give an example of R, S, T such that $R \circ (S \cap T) \neq (R \circ S) \cap (R \circ T)$.

14. The composition of two fuzzy relations R and S may be defined differently than in Subsection 6.1.3. Following, two definitions are cited, both for binary relations on finite universes U_1, U_2, U_3. Let R and S be binary relations on $U_1 \times U_2$ and $U_2 \times U_3$, respectively.

The max-probabilistic product composition of R and S is defined by

$$\{((u, w), \max_{v \in U_2} (\mu_R(u, v) \cdot \mu_S(v, w))) \mid u \in U_1, v \in U_2, w \in U_3\},$$

where $\cdot$ is the ordinary product of numbers (or probabilistic product from Exercise 6).

The max-bounded product composition of R and S is defined by

$\{((u, w), \max_{v \in U_2} (\max(0, \mu_R(u, v) + \mu_S(v, w)))) \mid u \in U_1, v \in U_2, w \in U_3\}.$

Show that the max-probabilistic product composition and the max-bounded product composition are associative and distributive over the union. (For definitions, see Exercise 6).

15. Determine the max-probabilistic product composition for fuzzy relations R and S from (a)–(f), Exercise 10.

16. Determine the max-bounded product composition for fuzzy relations R and S from (a)–(f), Exercise 10.

17. Let the fuzzy set TRUE over $U = [0, 1]$ be defined by the following S-function

$$\mu_{TRUE}(u) = S(u; 0.6, 0.7, 0.8),$$

for all $u \in [0, 1]$. Define

    a. NOT_TRUE,

    b. VERY TRUE,

    c. VERY VERY TRUE,

    d. FAIRLY TRUE.

18. For the following decision table:

	Attributes		Decisions		
	a	b	c	d	e
x_1	0	L	0	0	0
x_2	0	L	0	0	1
x_3	0	H	0	0	0
x_4	0	H	1	1	1
x_5	0	H	1	1	1
x_6	1	L	1	0	0
x_7	1	L	1	1	1

    a. Find certain and possible rules for decisions c, d, and e,

    b. Classify each of the partitions generated by $\{c\}$, $\{d\}$, and $\{e\}$, as $\{a, b\}$-definable or $\{a, b\}$-undefinable. For $\{a, b\}$-undefinable partitions, tell what their type (e.g., roughly $\{a, b\}$-definable, weak, and so on) is.

19. For the following decision table:

	Attributes			Decisions	
	a	b	c	d	e
x_1	0	0	L	L	0
x_2	0	0	H	L	1
x_3	0	1	H	L	2
x_4	0	1	H	R	3
x_5	1	0	H	S	3
x_6	1	0	H	S	4

find certain and possible rules for decisions d and e.

20. For the following decision table:

	Attributes			Decisions	
	a	b	c	d	e
x_1	0	0	R	A	0
x_2	0	1	S	B	1
x_3	0	2	T	C	2
x_4	1	0	T	B	3
x_5	1	1	R	C	4
x_6	1	1	R	C	5
x_7	1	2	S	A	6
x_8	1	2	S	A	7

find certain and possible rules for decisions d and e.

21. Let $A = (U, R)$ be an approximation space with $U = \{1, 2, 3, 4, 5, 6\}$.

 a. Find R and two subsets X and Y of U such that
$$\underline{R}(X \cup Y) \neq \underline{R}X \cup \underline{R}Y \text{ and } \overline{R}(X \cap Y) \neq \overline{R}X \cap \overline{R}Y.$$

 b. For sets X and Y from (a), compute qualities of lower and upper approximations of X and Y.

C H A P T E R
7
QUALITATIVE
APPROACHES

The basic principle of reasoning based on classical logic is that the reasoner's beliefs are true, and that the truth never changes. Thus the main need for reasoning is to add new beliefs to the current set of beliefs. In such a monotonic approach, common-sense reasoning, the frame problem, and the control problem, three problems studied by M. Minsky, J. McCarthy, and others (Doyle, 1979), cannot be resolved.

The first problem, common-sense reasoning, refers to the fact that in real life, people easily change their images of objects and processes, adjusting their beliefs to avoid errors. Old conclusions are abandoned and new evidence causes the addition of new beliefs. Briefly, reasoning based on common sense is non-monotonic.

The frame problem is one of the oldest problems of artificial intelligence. It refers to the fact that in real life, during any action most of the description of the state of a system will remain unchanged. Say that the system is a family room, with a TV set, a table, a number of armchairs, a sofa, and so on. The action is the family pet, a cat, entering the room through the window and leaving through the door. What is happening may be described by a sequence of snapshots, starting with the cat at the window and ending with the cat at the door. Most things in the family room remain untouched: They do not change their position, color, weight, and so on. It would be extremely difficult to describe what is unchanged. However, this must be done in a computer program because a computer cannot guess what happened: The situation must be explained.

The problem of control is how to make decisions taking current beliefs of the reasoner into account. The reasoner may even include his desires and intentions in the set of beliefs. The "inference rules" are rules of modification of the reasoner's beliefs. Many methods of nonmonotonicity refer to this problem.

181

7.1 Modal Logics

Modality is characterized by the presence of a modal functor (i.e., a non-truth-functional expression forming compound expressions out of simple ones). If F is a modal functor and A is a sentence, then the truth value of $F(A)$ is not entirely determined by the truth value of A.

Modality was explored in ancient and medieval logic as related to truth (i.e., as *alethic modality*). For example, *it is necessarily true that* A, *it is actually true that* A, *it is possibly true that* A. In 1951, G. H. Von Wright studied *epistemic modalities*, relating to knowledge, and *deontic modalities*, relating to duties. Examples of epistemic modalities are *it is known that* A, and *it is possible, for all that is known, that* A. Deontic modalities are *it is obligatory that* A, *it is permitted that* A, *it is forbidden that* A.

The modern history of modal logics starts from the work of C. I. Lewis in 1918, whose modal calculus was intended as a theory of strict implication. Modal logics include a series of systems of strict implication, denoted S1 to S5 (S stands for "strict").

Modal propositional calculus includes functors *it is necessary that*, denoted

$\Box$, and *it is possible that*, denoted $\Diamond$. The implication of an ordinary propositional calculus (see Table 2.1), and denoted $\rightarrow$, is also called a *material implication*, to distinguish it from *strict implication*, denoted $\Rightarrow$, and defined as follows:

$$P \Rightarrow Q \text{ if and only if } \neg\Diamond(P \wedge \neg Q),$$

where P and Q are propositions. In modal propositional calculus, the modal functors of necessity and possibility are prefixed to simple or compound propositions. Thus, they describe necessity or possibility of states of affairs and not properties of objects. An example of a modal propositional calculus, called M, T, S2', or the Feys/Von Wright system, axiomatized by R. Feys in 1937, is given here.

$$\Diamond P \text{ is defined as } \neg\Box\neg P,$$

$$P \Rightarrow Q \text{ as } \neg\Diamond(P \wedge \neg Q),$$

and

$$P \Leftrightarrow Q \text{ as } (P \Rightarrow Q) \wedge (Q \Rightarrow P),$$

where P and Q are propositions and $\Leftrightarrow$ denotes the *strict double implication*.

Axioms include those of classical propositional calculus and two additional ones:

1. $\Box P \rightarrow P$ (the reflexiveness axiom),

2. $\Box(P \rightarrow Q) \rightarrow (\Box P \rightarrow \Box Q)$.

Rules of inference:

1. Modus ponens for material implication $\rightarrow$,

2. Rule of substitution: From a proposition A in which a propositional variable V occurs at least once, infer whatever results from substituting any proposition B for V in A,

3. If P is a theorem (i.e., a proposition derived from axioms), infer $\Box P$ (*Gödel rule*, also called a *necessitation*).

Well-formed formulas are expressions that may be obtained by finitely many applications of the following rules:

1. A propositional variable is a well-formed formula,

2. If P and Q are well-formed formulas, so are $\neg P, P \vee Q, P \wedge Q, P \rightarrow Q,$ $P \leftrightarrow Q, \Box P, \Diamond P, P \Rightarrow Q,$ and $P \Leftrightarrow Q.$

The system S4 may be obtained from T by adding the axiom

$$\Box P \rightarrow \Box\Box P \text{ (the transitivity axiom)}.$$

The system S5 may be obtained from T by adding the axiom

$$\Diamond P \rightarrow \Box\Diamond P.$$

An example of a theorem that belongs to S3 (and therefore to S4), but not to S2 or to T is

$$(P \Rightarrow Q) \Rightarrow (\Box P \Rightarrow \Box Q).$$

Nevertheless, T includes the following theorems:

$$(P \Rightarrow Q) \Rightarrow (\Box P \rightarrow \Box Q),$$

and

$$\Box\Diamond P \rightarrow \Diamond P$$

see Marciszewski (1981).

From the additional axiom of S5, distinguishing it from T and from the last theorem, it follows that

$$\Box\Diamond P \leftrightarrow \Diamond P.$$

The preceding is an example of a reduction theorem. Additionally, in S5 there exist the following reduction theorems:

$$\Box\Box P \leftrightarrow \Box P,$$

$$\Diamond\Diamond P \leftrightarrow \Diamond P,$$

$$\Diamond\, \Box P \leftrightarrow \Box P,$$

$$\Diamond\, \Box\Diamond P \leftrightarrow \Diamond P.$$

In the system S5 every well-formed formula of the type MP, where M is the sequence of modal symbols and P is a well-formed formula, is equivalent to one of the following expressions:

$$\Box P, \Diamond P, \neg\Box P, \neg\Diamond P, \neg P, \text{ and } P.$$

Therefore, in the system S5, there are 6 nonequivalent modalities. It may be shown that S4 contains 14 such modalities and S3 contains 42 nonequivalent modalities.

An example of a modal predicate calculus is A. N. Prior's system, formulated on the basis of S5, with two rules for quantifiers, defined by J. Lukasiewicz in 1951. The system contains a system of classical propositional calculus. $\Diamond P$

is defined as $\neg\Box(\neg P)$, and three extra axioms are formulated:

1. $\Box(P \rightarrow Q) \rightarrow (\Box P \rightarrow \Box Q)$,

2. $\Box P \rightarrow P$,

3. $\neg\Box P \rightarrow \Box\neg\Box P$.

Gödel rule of inference is accepted:

if P is a theorem, infer $\Box P$,

and rules for quantifiers are

1. If $P \rightarrow Q$ is a theorem, so is $(\exists x)(P \rightarrow Q)$, provided x does not occur freely in Q,

2. If $P \rightarrow Q$ is a theorem, so is $P \rightarrow (\exists x)Q$.

The development of semantics for modal logic was done in the late fifties, with S. Kripke as the most influential contributor in 1959. S. Kripke's approach provides a set of assignments for each formula, replacing a unique assignment of classical predicate logic. These assignments are constructed as different states of affairs or *possible worlds*.

The term "possible world" is useful for many applications. For example, in a country having few political parties, it may be useful to consider different possible configurations of the government as a result of voting. A convenient way to describe the relevance of possible worlds is through a component of Kripke's structure, an *accessibility relation* R, defined on the set of possible worlds.

A pair $(W, W') \in R$ if and only if W' is possible with respect to W. The intuition is the following: If the current world is W, not every situation is possible from the point of view of W. The proposition P is necessarily true in a world W if and only if P is true in all worlds that are accessible from W. A proposition P is possibly true in a world W if and only if P is true in at least one world that is accessible from W. The properties of R determine the type of modal logic. When R is reflexive, it provides a structure for T. When R is reflexive and transitive, it is associated with S4, while when it is an equivalence relation, it describes S5.

7.2 Nonmonotonicity

Nonmonotonic reasoning is especially appropriate when the knowledge is incomplete, the universe of discourse is changing, and assumptions are temporary.

Nonmonotonic logics are systems in which the addition of new axioms can invalidate old theorems. Development of such logics is especially important for AI because they are suitable for studying common-sense reasoning, modeling rules with exceptions, revising beliefs after discovering new reasons, or taking defaults into account. An important initial contribution to the development of nonmonotonic logics was made in McDermott and Doyle (1980), and later in Moore (1984). R. C. Moore suggested a new approach to the problem and introduced autoepistemic logic. Another popular form of nonmonotonicity is default reasoning, in which plausible inferences are drawn from the current evidence without information to the contrary. This research was initiated by R. Reiter (1980).

Yet another nonmonotonic formalism is represented by circumscription, introduced by J. McCarthy in 1978. Circumscription is a form of nonmonotonic reasoning augmenting ordinary first-order logic; it formalizes arriving at the conclusion that the objects having specific properties are the only objects that have them.

An important foundation for implementation of a default reasoning is the truth maintenance system (Doyle, 1979), later generalized to the assumption-based truth maintenance system (de Kleer, 1986).

7.2.1 Nonmonotonic and Autoepistemic Logics

In monotonic logics, for two sets A and B of axioms, $A \subseteq B$ implies $\text{Th}(A) \subseteq \text{Th}(B)$, where $\text{Th}(A)$, $\text{Th}(B)$ denote theorem sets of A and B, respectively. This is no longer true in nonmonotonic logics.

D. McDermott and J. Doyle (1980) modify the classic first-order predicate calculus by introducing a propositional operator M whose meaning is *consistent with the theory*.

Due to some inexplicable phenomenon, an unaccountably popular example of nonmonotonic reasoning is a bird (a specific one called Tweety), which is (or is not) able to fly. The bird able to fly is described in the system of D. McDermott and J. Doyle by

$$(\forall x)(\text{bird}(x) \land M \text{ can\_fly}(x) \rightarrow \text{can\_fly}(x)),$$

or, informally, "for each x, if x is a bird and it is consistent to assert that x can fly, then x can fly".

As R. C. Moore (1985) points out, the system of D. McDermott and J. Doyle can have a single nonmonotonic inference rule with the following intuitive meaning:

"MP is derivable if $\neg P$ is not derivable".

The original logic defined in McDermott and Doyle (1980) gives such a weak notion of consistency that MP is not inconsistent with $\neg P$, i.e., P may be consistent with the theory and P may be false. D. McDermott attempted to support their nonmonotonic logic by modal calculi T, S3, and S5. As a result, nonmonotonic modal systems T, S4, and S5 were created. Unfortunately, the most plausible candidate for better formalizing the concept of consistency, nonmonotonic S5, collapsed to an ordinary S5.

As R. C Moore observed (Moore, 1985), D. McDermott and J. Doyle confused two different forms of nonmonotonic reasoning: default reasoning and autoepistemic reasoning. The preceding example of a bird is an example of default reasoning: In the absence of information to the contrary, the tentative conclusion is that a specific bird, Tweety, for example, can fly. However, "M can_fly (x)" from the same example may be interpreted differently: The only birds that cannot fly are the ones that can be inferred not to fly. Thus, the inference rule for M is now "MP is derivable if $\neg P$ is not derivable". This type of reasoning is not a form of default reasoning. Instead, it is reasoning about one's own knowledge or belief. R. C. Moore called it *autoepistemic reasoning*. Autoepistemic reasoning differs from default reasoning because the latter is defeasible, while the former is not. In default reasoning, conclusions are tentative and may be withdrawn with better information. This cannot happen in autoepistemic logic. For example, if we believe all cases of birds that cannot fly are known, the conclusion is final. Autoepistemic reasoning is nonmonotonic because the meaning of an autoepistemic statement is context-sensitive, dependent on the theory.

As an example, R. C. Moore considers two different theories. The first one consists of the following two axioms:

bird (Tweety),

$$(\forall x)(\text{bird}(x) \land M \text{ can\_fly}(x) \rightarrow \text{can\_fly}(x)).$$

MP means that P, i.e., can_fly (x), is consistent with that specific theory. Therefore, can_fly (Tweety) is a theorem of this theory. In the next theory, the only axioms are

¬can_fly(Tweety),

bird(Tweety),

$(\forall x)(\text{bird}(x) \wedge M \text{ can\_fly}(x) \rightarrow \text{can\_fly}(x))$.

In this theory, can_fly (Tweety) is not a theorem.

7.2.2 Default Logic

Default reasoning has been implemented in a number of programs, starting in 1972 with PLANNER, where the primitive THNOT as a goal succeeded if and only if its argument failed. Default reasoning is an economic way to fill slots in frames. Therefore, it is implemented in frame-based knowledge representation languages, such as KRL or FRL.

Default reasoning is frequently hidden under the name *closed-world assumption*. This concept is used in data bases where an unsuccessful search means nonexistence. For example, if, according to a timetable, there is no boat trip across Clinton Lake at 5:00 P.M. Sunday, the conclusion that such a boat trip does not exist is justified.

The example of a bird from 7.2.1 may be expressed as the following *default rule*:

$$\frac{\text{bird }(x): M \text{ can\_fly}(x)}{\text{can\_fly }(x),}$$

where *MP* means it is consistent to assume that *P*. The meaning is: "if *x* is a bird and it is consistent to assume that *x* can fly, infer that *x* can fly".

Formally, a *default theory*, $\Delta = (D, W)$ consists of a set *D* of defaults and a set *W* of well-formed formulas of a first-order predicate calculus. A default is any expression of the form

$$\frac{A: M B_1, M B_2,..., M B_m}{C},$$

where $A, B_1, B_2,..., B_m$, and *C* are well-formed formulas. *A* is called the *prerequisite*; $B_1, B_2,..., B_m$, the *justifications*; and *C*, the *consequent* of the default.

The default is intuitively interpreted as "if *A* is believed and each of B_1, $B_2,..., B_m$ can be consistently believed, then *C* may be believed". The set *W* is a set of axioms of Δ.

The simplest default is

$$\frac{: M B}{B}$$

that is, *A* is empty, $m = 1$, $B_1 = C = B$. The interpretation is: "if *B* is consistent with what is known, then it is believed".

Two types of defaults with $m = 1$ (i.e., with a single justification) are of special interest. A default with $m = 1$ and $B = C$ is called *normal*; a default with $m = 1$ and $B = C \wedge D$ for some D, is called *seminormal*. All defaults known in the literature are of one of these two types.

Defaults are used as nonmonotonic rules of inference together with those expressed in first-order logic. The set of all beliefs derivable from a default theory is called an *extension* of the default theory. Using the preceding interpretation of a default, W. Lukaszewicz (Lukaszewicz, 1984) showed that the existence of the extension of a default theory may depend on the order of invoking defaults. Therefore, he suggested another interpretation of a default rule: "If A is believed and it is consistent to believe $B_1, B_2,..., B_m$, then C may be believed under the assumption that it does not lead to inconsistency of justifications of the default, nor justifications of any other applied default".

The concept of an extension is crucial for default logic because an extension is a set of beliefs that is justified by what is known about a world. Default theories consisting of only normal defaults always have extensions. However, this is not always true for default theories consisting of seminormal defaults. The concept of an extension was modified by W. Lukaszewicz so that the existence of extensions is guaranteed (Lukaszewicz, 1984).

7.2.3 Circumscription

The motivations for circumscription come from an observation that in real life, frequently too many assumptions are necessary to make a decision or to place a belief. Circumscription is a kind of inference rule for computing a set of axioms. In *domain circumscription* (McCarthy, 1977, 1980), objects required by given information are all there are. In *predicate circumscription* (McCarthy, 1980), to circumscribe a well-formed formula A with respect to a predicate P contained in A is to state that the only objects satisfying P are those that have to on the basis of A. The domain circumscription is subsumed by the predicate circumscription. In 1984, J. McCarthy introduced *formula circumscription*, which is more general than predicate circumscription, an even more general *prioritized circumscription* (McCarthy, 1984). The formula circumscription uses a second-order logic.

Let P be a predicate symbol and let $A(P)$ be a well-formed formula of a first-order predicate calculus, containing P. $A(\Phi)$ means the result of replacing all occurrences of P in A by a well-formed formula Φ. The predicate circumscription of P in $A(P)$ is the following schema:

$$(A(\Phi) \wedge (\forall x)(\Phi(x) \to P(x))) \to (\forall x)(P(x) \to \Phi(x)),$$

where $x = (x_1, x_2,..., x_n)$, $n \geq 1$.

The circumscription of P in $A(P)$ may be interpreted in the following way: Objects described by P and necessary to satisfy $A(P)$ are the only ones that are necessary.

The proposition that can be obtained by circumscription of P in $A(P)$ is a result of *circumscriptive inference*. Circumscriptive inference is an example of nonmonotonic reasoning.

The following example is a modified example from McCarthy (1980). Let the predicate $P(x)$ be an isdwarf(x); let the well-formed formula $A(P)$ be

isdwarf(Bashful) $\wedge$ isdwarf(Doc) $\wedge$ isdwarf(Dopey) $\wedge$ isdwarf(Grumpy) $\wedge$
isdwarf(Happy) $\wedge$ isdwarf(Sleepy) $\wedge$ isdwarf(Sneezy).

Thus, Bashful, Doc, Dopey, Grumpy, Happy, Sleepy, and Sneezy are described by P and are necessary to satisfy $A(P)$. Circumscribing isdwarf in the preceding formula $A(P)$ yields

$(\Phi(\text{Bashful}) \wedge \Phi(\text{Doc}) \wedge \Phi(\text{Dopey}) \wedge \Phi(\text{Grumpy}) \wedge \Phi(\text{Happy}) \wedge \Phi(\text{Sleepy}) \wedge$
$\Phi(\text{Sneezy}) \wedge (\forall x)(\Phi(x) \rightarrow \text{isdwarf}(x))) \rightarrow (\forall x)(\text{isdwarf}(x) \rightarrow \Phi(x))$

Say that $\Phi(x)$ is the following formula:

$(x = \text{Bashful}) \vee (x = \text{Doc}) \vee (x = \text{Dopey}) \vee (x = \text{Grumpy}) \vee (x = \text{Happy}) \vee$
$(x = \text{Sneezy}) \vee (x = \text{Sleepy}).$

Then the left side of the above implication is true. Therefore,

$$(\forall x)(\text{isdwarf}(x) \rightarrow \Phi(x)).$$

Thus, the conclusion is that the only dwarfs are Bashful, Doc, Dopey, Grumpy, Happy, Sleepy, and Sneezy. The circumscription inference from the preceding example is nonmonotonic, since if isdwarf(Krasnal) is adjoined to $A(P)$, the example's conclusion is no longer true.

7.2.4 Truth Maintenance System

One of the main contributions to the development of truth maintenance systems (known under the abbreviation TMS and also called *belief revision* or *reason maintenance systems*) was done by J. Doyle (1979). The system, presented by J. Doyle, is justification-based, with two main components: a problem solver, which draws inferences in the form of justifications, and a truth maintenance system, which records justifications, makes nonmonotonic inferences, and checks for contradictions (see Figure 7.1). The problem solver is concerned with the knowledge of the domain. The truth maintenance system uses a constraint satisfaction procedure, called truth maintenance, to determine the set of beliefs, thus producing nonmonotonic inferences. Contradictions are not tolerated by the truth maintenance system. They are removed by inserting additional justifications in the special procedure called dependency-directed backtracking. Therefore, the data are always consistent.

The original truth maintenance system was later extended to an assumption-based truth maintenance system, or briefly, ATMS, by J. de Kleer in 1986. Instead of relying solely on justifications, this extended system is also based on manipulating assumptions. Inconsistencies are not exterminated anymore, and

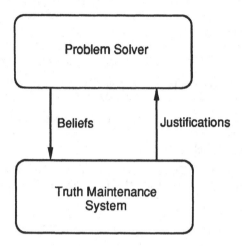

Figure 7.1 Truth Maintenance System

most backtracking is avoided. This system, together with some additional features, was incorporated into the reasoning tools ART and KEEworlds. Note that the assumption-based truth maintenance system was further generalized into a clause management system by R. Reiter and J. de Kleer (Reiter and de Kleer, 1987).

In the truth maintenance system, a special data structure, a *node*, represents the belief in a problem-solver piece of knowledge, such as a fact or inference rule. Relations on the set of all nodes are represented by *justifications*. Each node has a *support status* associated with it. Such a node may have many justifications, among them a few valid ones. One of these is selected as the current supporting justification and called simply a *supporting justification*. A node without any currently valid justification is *out* of the current set of beliefs. The set of justifications associated with a node is called a *justification set*. The problem solver may insert or delete nodes and justifications, or it may mark nodes as contradictory.

A new justification for a node may cause the truth maintenance system to classify the node as being *in* (the set of beliefs). This may, in turn, allow other nodes to be *in* (the set of beliefs), as a result of invoking the truth maintenance procedure.

A contradiction occurs when a node marked as contradictory is *in*. Then the dependency-directed backtracking determines assumptions, causing the contradiction. Additional justifications are added to the *culprits* until the contradiction is *out* of the set of beliefs.

The truth maintenance system determines an assignment of belief status for all nodes. The main philosophical idea of the truth maintenance system (Doyle,

1979) is that the current set of beliefs and desires arises from the current set of reasons for beliefs and desires. One of the consequences of this point of view is that the main emphasis is on justified belief, ignoring the question of truth.

7.2.4.1   Justifications

The truth maintenance system uses two different forms of justifications (or reasons) for belief. The first one is called *support-list justification*. It has the following format:

$$(SL \ (inlist) \ (outlist)),$$

where *inlist* and *outlist* are two lists of nodes. The support-list justification is *valid* if and only if each node in its *inlist* is *in* and each node in its *outlist* is *out*. A belief in a node is based not only on current beliefs in other nodes but also on the lack of current belief in other nodes. Thus, justifications of the truth maintenance system are nonmonotonic. When both lists are empty, the support-list justification represents *premise* and is always valid. Therefore, the node it justifies will always be *in*. A support-list justification with a nonempty list and empty *outlist* represents a monotonic argument. *Assumptions* are defined as nodes whose supporting justifications have nonempty *outlists*. Therefore, assumptions are nonmonotonic justifications.

The second type of justification is called *conditional-proof*. Its format is

$$(CP \ (consequent) \ (inhypotheses) \ (outhypotheses))$$

where *consequent* means consequent of a node, and *inhypotheses* and *outhypotheses* describe the validity of certain hypothetical pros and cons. A conditional-proof justification is *valid* if and only if the consequent node is *in* each node of the *inhypotheses*, and each node of the *outhypotheses* is *out*. If the set of all *outhypotheses* is empty, a node justified by a conditional-proof justification represents the implication whose antecedents are the *inhypotheses* and whose consequent is the consequent of the conditional-proof justification.

The truth maintenance system deals practically exclusively with the support-list justifications, since each valid conditional-proof justification is transformed into an equivalent support-list justification by the system.

7.2.4.2   Node Types

Let node X be *in* the current set of beliefs. This is equivalent to saying that the truth maintenance system has already selected one current support-list justification for X, called *supporting justification*, and that this justification is valid. The *supporting-node set* of X is the set of all nodes listed in the *inlist* and the *outlist* or its supporting justification.

Let node X be *out* of the current set of beliefs. In this case, the *supporting-node set* of X is the set of nodes, such that exactly one node is selected from each justification in the justification set of X, either an *outnode* from the *inlist* or an *innode* from the *outlist*. The node X cannot change its support status into *in* unless at least one of its supporting nodes changes its support status, or a new valid justification will be added to the justification set of X.

The *antecedent set* of a node X is the supporting-node set of X if X is *in*. When X is *out*, it has no antecedents.

The *foundation set* of a node is the transitive closure of the antecedent set of the node (i.e., the union of the set of all the antecedents of the node, the set of all the antecedents of the antecedents of the node, and so on). The *ancestor set* of a node is the transitive closure of the supporting-node set of the node (i.e., the union of the set of all the supporting nodes of the node, the set of all the supporting nodes of the supporting sets of the node, and so on). The ancestor set is the set of all nodes that may affect the support status of the node.

The *consequence set* of a node X is the set of all nodes that include X in at least one justification from their justification set. The *affected-consequence set* of a node X is a subset of the consequence set of X, such that the subset consists of all consequences that contain X in their supporting-node sets. The *believed-consequence set* of a node X is a subset of the consequence set of X, such that the subset consists of all *in* consequences that contain X in their antecedent sets.

The *repercussion set* of a node is the transitive closure of the affected-consequence set of the node (i.e., the union of the set of all the affected consequences of the node, the set of all the affected consequences of the affected consequences of the node, and so on). The *believed-repercussion set* of a node is the transitive closure of the believed-consequence set of the node (i.e., the union of the set of all the believed consequences of the node, the set of all the believed consequences of the believed consequences of the node, and so on).

The preceding definitions are illustrated by an example of a system from

Table 7.1 A Truth Maintenance System

Node	Justification Set
A	{J1 = (SL () ())}
B	{J2 = (SL (A) (C))}
C	{J3 = (SL (A) (B))}
D	{J4 = (SL (A) (E)}
E	{J5 = (SL (A) (D))}
F	{J6 = (SL (C) ()), J7 = (SL (D) ()), J8 = (SL () (E))}
G	{J9 = (SL (B) ()), J10 = (SL (E) ()), J11 = (SL () (D))}

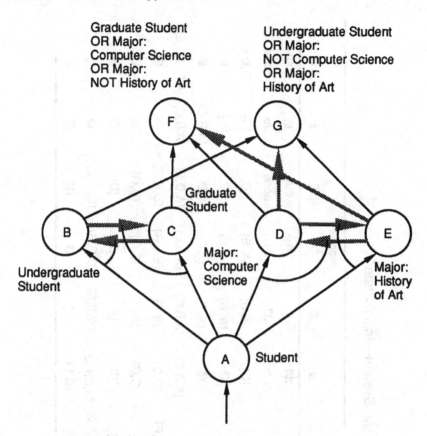

Graduate Student
OR Major:
Computer Science
OR Major:
NOT History of Art

Undergraduate Student
OR Major:
NOT Computer Science
OR Major:
History of Art

Figure 7.2 System from Table 7.1

Table 7.1. The system is presented in Figure 7.2. Ordinary arcs represent *in*lists; gray arcs represent *out*lists. A justification whose union of *in*list and *out*list consists of more than a single member is represented in Figure 7.2 by a set of directed arcs; each arc corresponds to a member of either *in*list or *out*list, and all such arcs are embraced by an additional undirected arc. Node set types for all nodes of the system are listed in Table 7.2. The system contains information about the status of a person. The person is a student because justification *J*1 for that piece of information is always valid. The current set of beliefs is that the person is a graduate student in Computer Science.

7.2.4.3 Circular Arguments

Let us say that the problem solver will add justification (SL (*F*) ()) to the justification set of node *C* of the system from Table 7.1. Node *F* is justified by (SL (*C*) ()). Thus, the support statuses of both *C* and *F* are *in*. Then both nodes, *C*

Table 7.2 Node Set Types of the System—Initial State

Node Set Types		A	B	C	D	E	F	G
	Nodes							
Support Status		in	out	in	in	out	in	out
Supporting Justification		J1	none	J3	J4	none	J6	none
Supporting-Node Set		∅	{C}	{A, B}	{A, E}	{D}	{C}	{B,D,E}
Antecedent Set		∅	∅	{A, B}	{A, E}	∅	{C}	∅
Foundation Set		∅	∅	{A,B}	{A,E}	∅	{A,B,C}	∅
Ancestor Set		∅	{A,B,C}	{A,B,C}	{A,D,E}	{A,D,E}	{A,B,C}	{A,B,C,D,E}
Consequence Set		{B,C,D,E}	{C,G}	{B,F}	{E,F,G}	{D,F,G}	∅	∅
Affected-Consequence Set		{C,D}	{C,G}	{B,F}	{E,G}	{D,G}	∅	∅
Believed-Consequence Set		{C,D}	{C}	{F}	∅	{D}	∅	∅
Repercussion Set		{B,C,D,E,F,G}	{B,C,F,G}	{B,C,F,G}	{D,E,G}	{D,E,G}	∅	∅
Believed-Repercussion Set		{C,D,F}	{C,F}	{F}	∅	{D}	∅	∅

and F, should remain *in* because their justifications form circular arguments for C and F in terms of each other.

Another example of circularity includes two nodes of the system from Table 7.1, namely B and C. The justification for B is (SL (A) (C)); for C, it is (SL (A) (B)). As long as A is *in*, if B is *in*, C must be *out* and vice versa.

In yet another form of circularity, called *unsatisfiable*, the support status of a node X can be neither *in* nor *out*. For example, a supporting justification for X is (SL $()$ (X)).

The algorithms of the truth maintenance system must guarantee that no node is believed because of a circular argument.

7.2.4.4 Truth Maintenance

The truth maintenance system determines the assignment of support statuses for all nodes. Deleting a justification, inserting an invalid one, or inserting a valid justification to a node X whose support status is *in* does not cause a problem. The only essential problem is when a new valid justification is added to a node X that is *out*. In this case, the node and its repercussions must be updated. Subsequently, guidelines for the procedure of truth maintenance, a case of inserting a new justification, are sketched.

Insert the new justification into the node's justification set. Add node X to the set of consequences of each of the nodes mentioned in the justification. If node X is *in*, the procedure stops. If node X is *out*, the justification must be checked for validity. If it is invalid, add an *out*node from its *in*list or an *in*node from its *out*list to the set of supporting nodes of node X. If it is valid, go to the next step because updating beliefs is required.

Check the affected-consequence set of node X. If the set is empty, change the support status of node X to *in* and change the supporting-node set of X to the union of *in*list and *out*list; then halt. If the affected-consequence set of X is nonempty, make a list L containing X and all its repercussions, marking the supporting statuses for all members of L as *nil*.

Before the next step is described, two additional definitions should be cited. A support-list justification is *well-founded valid* if and only if each node in its *in*list is *in* and each node in it *out*list is *out*. A support-list justification is *well-founded invalid* if and only if a node exists in its *in*list that is *out* or a node exists in its *out*list that is *in*. A support-list justification is *not well-founded* if and only if each node in its *in*list is *in* and no node in its *out*list is *in*; otherwise, it is *not well-founded invalid*.

For each of the members of L, execute the following subprocedure: If the support status of a node is *in* or *out*, do nothing. Otherwise, inspect all justifications from the supporting-node set, checking them for well-founded validity until a well-founded valid justification is discovered or the entire supporting set is inspected.

If a well-founded valid justification is found, then the justification has the status of supporting, the supporting-node set is recreated, the new one is the

Table 7.3 Node Set Types of the System—New State

Node Set Types			Nodes				
	A	B	C	D	E	F	G
Support Status	in	in	out	in	out	in	in
Supporting Justification	J1	J12	none	J4	none	J7	J9
Supporting-Node Set	∅	∅	{B}	{A,E}	{D}	{D}	{B}
Antecedent Set	∅	∅	∅	{A,E}	∅	{D}	{B}
Foundation Set	∅	∅	∅	{A,E}	∅	{A,D,E}	{B}
Ancestor Set	∅	∅	{B}	{A,D,E}	{A,D,E}	{A,D,E}	{B}
Consequence Set	{A,B,C,D,E}	{C,G}	{B,F}	{E,F,G}	{D,F,G}	∅	∅
Affected-Consequence Set	{D}	{C,G}	∅	{E,F}	{D}	∅	∅
Believed-Consequence Set	{D}	{G}	∅	{F}	{D}	∅	∅
Repercussion Set	{D,E,F}	{C,G}	∅	{D,E,F}	{D,E,F}	∅	∅
Believed-Repercussion Set	{D,F}	{G}	∅	{F}	{D,F}	∅	∅

union of the *in*list and the *out*list, the support status of the node is *in*, and the subprocedure is recursively performed for all consequences of the node with the status *nil*. If the supporting-node set consists of only well-founded invalid justifications, the node is marked *out*, its supporting-node set is recreated (the new one consisting of either an *out*node from the *in*list or an *in*node from the *out*list for each justification from the node's justification set), and the subprocedure is recursively performed for all *nil*-marked consequences of the node. At this point, some nodes from *L* may still have *nil* marks due to circularities.

If all nodes in *L* have the supporting status *in* or *out*, skip this step. Otherwise, for each node marked *nil*, inspect all justifications from its supporting-node set for not well-founded validity until a not well-founded valid justification is discovered or the entire supporting-node set is inspected. If a not well-founded valid justification is found, then check if the affected-consequence set of the node is empty. If it is, then the justification has the status of supporting, the supporting-node set is recreated as previously described, the support status of the node is *in*, and the step is recursively performed for all consequences of the node with the status *nil*.

If the affected-consequence set is nonempty, then all members of the set and the node are remarked with *nil* and reexamined by this step. If the supporting-node set consists of only not well-founded invalid justifications, the node is marked *out*, its supporting-node set is recreated as previously described, and the step is recursively performed for all *nil*-marked consequences of the node.

After this step, all nodes have support status *in* or *out*. Then the truth-maintenance system searches for *in*nodes marked as contradictions. For each such node, the dependency-directed backtracking is called. Finally, all changes in node-support statuses are reported to the problem solver.

For example, the justification $J12 = (SL\ ()\ ())$ is inserted into the justification set of node *B* of the system represented by Table 7.1. Both the *in*list and the *out*list of $J12$ are empty, so no additional insertions into the sets of consequences are necessary. The justification is valid, and the node *B* is *out*.

The affected-consequence set of node *B* is nonempty; hence, list *L* is created: $L = \{B, C, F, G\}$. Temporarily, all nodes from *L* are marked *nil*. There exists the well-founded valid justification $J12$ for node B, and thus $J12$ becomes the supporting justification. Thus, the support status of *B* is *in* and the supporting-node set is the empty set. Note that $J2$ is not well-founded because the current support status of *C* is *nil*. The next candidate from *L* is *C*. Its justification, $J3$, is well-founded invalid; thus, the support status of *C* is *out*, and the supporting-node set of *C* is $\{B\}$. The same algorithm step processes nodes *F* and *G*, successfully searching for well-founded justifications $J7$ and $J9$, respectively. The new state of the system is described by Table 7.3.

7.2.4.5   Default Assumptions

If the default value is selected from two alternatives, the default node may be justified nonmonotonically on the basis of the other node being *out*. For example, in the system from Table 7.1, the node C, representing a graduate student, is justified by (SL (A) (B)), where the node B is interpreted as an undergraduate student.

For a more general case of n alternatives $A_1, A_2,..., A_n$, let G denote a node that represents a reason for an assumption to choose the default. Say that A_i is chosen to be a default. Then A_i should be justified with

$$\text{(SL } (G) \ (A_1 A_2 \cdots A_{i-1} A_{i+1} \cdots A_n)).$$

When no additional information is available, the remaining nodes should have no valid justifications. Thus, A_i will be *in* and all remaining nodes will be *out*. When some valid justification is added to an alternative node $A_j, j \neq i$, then A_j will become *in* and A_i will become *out*.

In some cases, the alternatives $A_1, A_2,..., A_n$ are linearly ordered, and they should be tried in that order. Let G denote a node that represents a reason for trying alternatives in the order $A_1, A_2,..., A_n$. Then each should be justified with

$$\text{(SL } (G \neg A_{i-1}) \ (\neg A_{i+1})),$$

where $i = 1, 2,..., n - 1$; the node A_n should be justified with

$$\text{(SL } (G \neg A_i) \ ()).$$

The problem solver may reject alternative A_1 by justifying alternative $\neg A_1$. This will change the support status of A_1 and A_2 from *in* to *out* and from *out* to

Table 7.4 A Truth Maintenance System

Node	Justification Set
G	{a valid justification}
A_1	{(SL $(G) \ (\neg A_1)$)}
$\neg A_1$	
A_2	{(SL $(\neg A_1) \ (\neg A_3)$)}
$\neg A_2$	
A_3	{(SL $(\neg A_2) \ (\neg A_4)$)}
$\neg A_3$	
A_4	{(SL $(\neg A_3) \ ()$)}
$\neg A_4$	

in, respectively. In turn, A_2 may be rejected by justifying $\neg A_2$, and so on.

For example, if G is a node that represents a reason for trying the alternatives in the order A_1, A_2, A_3, A_4, then Table 7.4 presents node justifications.

7.2.5 Reasoned Assumptions

The idea of reasoned assumptions is presented in Doyle (1985). Related concepts, like the UNLESS operator, were studied by E. Sandewall in 1972; and *censored rules* in *variable precision logic* were studied by R. S. Michalski and P. H. Winston (Michalski and Winston, 1986).

A *simple reason* is a rule of the type "*A* without *B* gives *C*", denoted

$$A \parallel B \Vdash C.$$

If A, B, C are potential beliefs, then the meaning of the simple reason is that items in C should be believed if all items in A are believed and none of those in B are believed. Conclusions drawn from simple reasons are called *reasoned assumptions* (reasoned because they are derived from reasons, assumptions because they depend on the lack of belief in items in B). The Tweety example is now formulated as

{Bel(Tweety is a bird)} ‖ {Bel(Tweety cannot fly)} ⊩ {Bel(Tweety can fly)}.

7.3 Plausible Reasoning

Plausible reasoning was invented by N. Rescher (1976). The theory presents an approach to managing conflicts in data. It is assumed that data are supported by unreliable sources of different degrees of plausibility. Plausible reasoning offers guidelines on how to reason in such a situation.

The basic concept is that of a nonempty set $S = \{P_1, P_2, ..., P_n\}$ of *plausible propositions*. Each such proposition is supported by a source (of information) of some positive *degree of reliability*. The greater the reliability of the source, the greater the plausibility of a proposition. As a result, a *degree of plausibility* is assigned to each proposition $P \in S$, a number between 0 and 1. The degree of plausibility for P is denoted $/P/$. For P, $/P/ = 1$ if and only if the proposition P is certain, i.e., true. If $/P/ = 0$, the proposition P is not plausible at all. All certain propositions from S must be mutually consistent. The set S need not contain all derivable propositions. Conflicts may occur among the elements of S. Rescher offers a number of rules of manipulating plausibility. These rules are cited and illustrated by examples here.

If more than one source supports a proposition P, then the source of the greatest degree of reliability determines the degree of plausibility of P. For example, source B from Table 7.5 supports $P \wedge Q$ with degree of reliability equal to 0.8, and source C supports the same proposition with degree of reliability equal to 0.7. Thus, the degree of plausibility for $P \wedge Q$ is 0.8.

Table 7.5 Degrees of Reliability

Source	Degree of Reliability	Supported Propositions
A	0.9	$P \rightarrow R$
B	0.8	$P \wedge Q, P \rightarrow R$
C	0.7	$P, P \wedge Q$

Set S may be augmented by propositions derived from members of S (i.e., the consequences of S). For example, if $S = \{P, P \wedge Q, P \rightarrow R\}$, with the plausibility assignment from Table 7.6, then the augmented set is $\{P, P \wedge Q, P \rightarrow R, R\}$, since R is derived from P and $P \rightarrow R$.

The degree of plausibility of a proposition that is derivable from some consistent subset T of the set of plausible propositions S is not less than the least degree of plausibility from T. For example, in the set $S = \{P, P \wedge Q, P \rightarrow R, R\}$, with the plausibility assignment from Table 7.6, $/R/ \geq 0.7$ because $/Q/ = 0.7$ and $/P \rightarrow R/ = 0.9$. The same rule holds when data are inconsistent. For example, the following degrees of plausibility follow from Table 7.7: $/\neg P \vee Q/ = 0.9$, $/P/ = 0.6$, $/Q/ = 0.6$, and $/\neg Q/ = 0.9$. Moreover, $\neg P \vee Q$ and P entails Q, $/Q/ \geq \min(0.9, 0.6) = 0.6$; similarly, $\neg P \vee Q$ and $\neg Q$ entails $\neg P$, $/\neg P/ \geq \min(0.9, 0.6) = 0.9$.

In the example, $/P/ = 0.6$ and $/\neg P/ = 0.9$. There is nothing wrong with this, as the degree of plausibility is not a probability. Plausibility, as understood by Rescher, is a classificatory concept. The numbers are useful as representatives of some degrees of plausibility and may be replaced by a linearly ordered set of names such as highly-plausible, medium-plausible, only-somewhat-plausible, and so on.

Table 7.6 Degrees of Plausibility

Propositions	Degree of Plausibility
P	0.7
$P \wedge Q$	0.8
$P \rightarrow R$	0.9

Table 7.7 Degrees of Reliability

Source	Degree of Reliability	Supported Propositions
A	0.9	$\neg P \vee Q, \neg Q$
B	0.6	P, Q

It may happen that a proposition P may be a consequent of the set S of plausible propositions in a number of ways. Let

$$P_1^1, P_2^1, ..., P_{n_1}^1,$$

$$P_1^2, P_2^2, ..., P_{n_2}^2,$$

$$...........$$

$$P_1^i, P_2^i, ..., P_{n_i}^i$$

be different sequences of elements of S; each row entails P. Then

$$/P/ \geq \max_i \min_j P_j^i.$$

For example, for $S = \{\neg P, \neg P \to Q, P \vee Q\}$, and $/\neg P/ = 0.7$, $/\neg P \to Q/ = 0.5$, and $/P \vee Q/ = 0.7$, there exist two different sequences of elements on S, both entailing Q:

$$\neg P, \neg P \to Q$$

and

$$\neg P, P \vee Q.$$

Therefore,

$$/Q/ \geq \max (\min (/\neg P/, /\neg P \to Q/), \min (/\neg P/, /P \vee Q/))$$

$$= \max (\min (0.8, 0.5), \min (0.8, 0.7))$$

$$= \max (0.5, 0.7)$$

$$= 0.7.$$

Augmenting the set S of inconsistent plausible propositions may cause need for revising the original degrees of plausibility for members of S even if only consequents are added. For example, let S be a set $\{P, Q, \neg Q\}$, and $/P/ = 0.6$, $/Q/ = 0.7$, and $/\neg Q/ = 0.8$. The proposition Q entails $P \vee Q$; hence, $/P \vee Q/ \geq 0.7$. Then $\neg Q$ and $P \vee Q$ entail P; thus, $/P/ = 0.6 \geq \min (/\neg Q/ = 0.8, /P \vee Q/),$

or $/P \vee Q/ \leq 0.6$, a contradiction. Therefore, the original value for $/P/$ should be reset.

The *neutralization rule* says that if the set S of plausible propositions is augmented to a set S^+ of plausible propositions, and if in S^+ there exists a consistent subset T such that a proposition P may be derived from T, which is "clearly untenable", then S^+ should contain a proposition P with such a high degree of plausibility that it gives it an advantage over P. Information whether P is "clearly untenable" is given by an expert. For example, S is the following set of plausible propositions:

{Tweety can fly, ¬(Tweety is a penguin), ¬(Tweety is an ostrich),
(Tweety is a penguin) ∨ (Tweety is an ostrich)},

and a highly reliable source is supporting (Tweety is a penguin) ∨ (Tweety is an ostrich). Then "Tweety can fly" is clearly untenable, so the proposition ¬(Tweety can fly) with high degree of plausibility should be added to S.

One of the main tasks of plausibility reasoning is conversion of inconsistent sets of plausible propositions into consistent ones. Let S be a set of plausible propositions, consistent or not. A nonempty subset T of S is called a *maximal consistent subset of* S if and only if it is consistent and no element of $S - T$ can be added to T without causing an inconsistency.

For example, if $S = \{P, Q, \neg Q, P \vee Q\}$, then two maximal consistent subsets of S exist : $\{P, Q, P \vee Q\}$ and $\{P, \neg Q, P \vee Q\}$. Among such maximal consistent subsets, some may be more desirable because of their plausibility. Therefore, the process of plausible screening was introduced by N. Rescher to select *plausibilistic consequences*. If a maximal consistent subset T is determined by the rejection of any proposition of high plausibility, then such a T is eliminated as the plausibilistic consequence. For example, if in the preceding example $/P/ = 0.8$, $/Q/ = 0.2$, $/\neg Q/ = 0.9$, and $/P \vee Q/ = 0.6$, then $\{P, \neg Q, P \vee Q\}$ is the plausibilistic consequence.

7.4 Heuristic Methods

In this section, two heuristic approaches will be discussed, both developed in the eighties.

7.4.1 Endorsements

A heuristic reasoning about uncertainty was introduced by P. R. Cohen and M. R. Grinberg (1983). In this method, a set of context-dependent rules controls the records of sources of uncertainty, called *endorsements*. Information about uncertainty (e.g., whether the evidence is for or against a proposition and the strength of the evidence), is represented directly by endorsements. The method was used in plan-recognition program HMMM, recognizing the intention of a user to adapt a few plans, taking into account the user's previous actions.

The idea can be illustrated by an example taken from (Sullivan and Cohen, 1985). Two plans are considered, plan 1 and plan 2, defined by sequences of three steps: *abc* and *bde*, respectively. The user, however, selected the sequence *abd*. The first step of the user's sequence, the letter *a*, indicates that plan 1 was chosen. The second step, *b*, is further evidence that plan 1 was chosen. The third step, *d*, indicates that perhaps the first letter was a mistake and the user wants to implement plan 2. The program HMMM uses endorsements to execute this kind of reasoning (see Table 7.8). Some endorsements are *positive*, supporting the interpretation with which they are associated; some are *negative*, not believing the associated interpretations.

In order to combine endorsements from previous steps with the current step endorsement, some heuristic combining rules are used:

(1) If (plan *i*: *x* could be a mistake, –) and (plan *i*: *xy* continuity is desirable, +) then erase (plan *i*: *x* could be a mistake, –),

(2) If (plan *i*: *xy* continuity is undesirable, –) and
(plan *i*: *yz* continuity is desirable, +) and
(plan *i*: *y* is another possibility, –) and
(plan *j*: *y* is another possibility , –),
then erase (plan *i*: *xy* continuity is undesirable, –).

Table 7.8 Endorsements of Two Plans

Step	Interpretation	Endorsements
a	the first step of plan 1	(plan 1: *a* is the only possibility, +)
		(plan 1: *a* could be a mistake, –)
b	the second step of plan 1	(plan 1: *ab* continuity is desirable, +)
		(plan 1: *b* is another possibility, –)
		(plan 1: *b* could be a mistake, –)
b	the first step of plan 2	(plan 2: *ab* continuity is undesirable, –)
		(plan 2: *b* is another possibility, –)
		(plan 2: *b* could be a mistake, –)
d	the second step of plan 2	(plan 2: *d* is the only possibility, +)
		(plan 2: *bd* continuity is desirable, +)
		(plan 2: *d* could be a mistake, –)

Table 7.9 Combined Endorsements

Step	Interpretation	Endorsements
a	the first step of plan 1	(plan 1: *a* is the only possibility, +)
b	the second step of plan 1	(plan 1: *ab* continuity is desirable, +)
		(plan 1: *b* could be a mistake, −)
b	the first step of plan 2	(plan 2: *b* is another possibility, −)
d	the second step of plan 2	(plan 2: *d* is the only possibility, +)
		(plan 2: *bd* continuity is desirable, +)
		(plan 2: *d* could be a mistake, −)

In the example, after using the preceding rules, the combined endorsements are presented in Table 7.9.

Some *numerical weights* are used to represent the strengths of endorsements. These weights are adjusted as a result of combination of endorsements. As the authors of the endorsement approach admit (Sullivan and Cohen, 1985), their method is subjective and domain-dependent, as are the heuristics they are using.

7.4.2 CORE

A system CORE of plausible reasoning, based partly on R. S. Michalski's variable-valued logic calculus (Michalski, 1975) and partly on human plausible reasoning (Carbonell and Collins, 1973), (Collins, 1978), was initiated in Collins and Michalski (1986).

The system is highly heuristic and contains many concepts grouped into primitives (such as expressions, arguments, descriptors, terms, references, statements, dependencies between terms, implication between statements), operators (generalization, specialization, similarity, dissimilarity), certainty operators (likelihoods for implications, degrees of certainty, typicality, similarity, frequencies of reference, dominances), and transforms (argument-based and reference-based). The basic assumption is that human knowledge is represented in *structures* (dynamic hierarchies), interconnected by *traces*. Inference is done mainly through argument-based and reference-based *transforms* on statements, and also by other patterns, such as derivation from mutual implication or mutual dependency.

7.5 Concluding Remarks

Qualitative approaches to uncertainty are well suited for handling incompleteness of information. Among qualitative approaches, nonmonotonicity is a very popular topic. Relationships between different models of nonmonotonicity have been studied, for example, between default logic and circumscription (Grosof, 1984; Imielinski, 1985; Etherington, 1987) or between default and autoepistemic logics (Konolige, 1987).

The next version of the truth maintenance system is an assumption-based truth maintenance system, briefly ATMS, introduced by de Kleer (1986). ATMS maintains an entire collection of assumption sets and is able to work with inconsistent information.

A number of attempts have been made to add numerical measures of uncertainty to default logic, initiated by E. Rich (1983). M. L. Ginsberg (1984, 1985) used Dempster-Shafer theory for this purpose, while R. Yager (1987) and D. Dubois and H. Prade (1988b) used possibility theory. There are attempts to enhance ATMS with numerical values representing uncertainty, for example, B. D'Ambrosio (1987) or K. B. Laskey and P. E. Lehner (1988).

K. Konolige introduced a deduction model, developed within quantified modal logic, which is a new formal model of belief (1986). N. J. Nilsson introduced probabilistic logic, in which truth values of propositions are probabilities (1986). However, the resulting theory is monotonic.

Until recently, the AI community paid little attention to temporal logic. In temporal logic, the same proposition may have a different truth value at different times. Recently, temporal logic has become quite popular. For example, Y. Shoham provided a logic called chronological ignorance, on the basis of modal and temporal logics. Chronological ignorance may be useful in common-sense reasoning (Shoham, 1988).

Plausibility reasoning, like rough set theory, seems to be a good technique for preprocessing knowledge before it is used by an expert system, when plausible consequences are drawn from inconsistent facts. The CORE method was recently expanded, modified, and implemented as the APPLAUS system (Dontas and Zemankova, 1988).

Exercises

1. Show that the following are theorems of T calculus:

    a. $\Box(P \Rightarrow P)$,

    b. $(P \wedge \neg P) \Rightarrow Q$,

    c. $Q \Rightarrow (P \vee \neg P)$,

    d. $\Box(P \wedge Q) \Leftrightarrow (\Box P \wedge \Box Q)$,

    e. $\Diamond(P \vee Q) \Leftrightarrow (\Diamond P \vee \Diamond Q)$,

   f. $(\Box P \vee \Box Q) \Rightarrow \Box(P \vee Q)$,

   g. $\Diamond(P \wedge Q) \Rightarrow (\Diamond P \wedge \Diamond Q)$.

2. Determine sets of objects by circumscriptive inference for a predicate P isdwarf and the following well-formed formulas $A(P)$:

   a. isdwarf(Bashful) $\vee$ isdwarf(Doc) $\vee$ isdwarf(Dopey) $\vee$ isdwarf(Grumpy) $\vee$ isdwarf(Happy) $\vee$ isdwarf(Sleepy) $\vee$ isdwarf(Sneezy),

   b. (isdwarf(Bashful) $\wedge$ isdwarf(Doc) $\wedge$ isdwarf(Dopey)) $\vee$ (isdwarf(Grumpy) $\wedge$ isdwarf(Happy)) $\vee$ (isdwarf(Sleepy) $\wedge$ isdwarf(Sneezy)).

3. Say that the only blackbird you know is Sweety. Therefore, seeing a blackbird, your assumption is that it is Sweety. Express this formally by predicate circumscription. Explain why your reasoning is nonmonotonic.

4. Modify the system from Table 7.1 so that node F will represent a graduate student majoring in computer science and not history of art, and node G will represent an undergraduate student majoring not in computer science but history of art. Determine justification for all nodes (i.e., modify Table 7.1). Then, assuming the same support statuses as in Table 7.2, determine the supporting justifications, supporting-node sets, antecedent sets, foundation sets, ancestor sets, consequence sets, affected-consequence sets, believed-consequence sets, repercussion sets, and believed-repercussion sets for all nodes (i.e., modify Table 7.2).

   a. run the truth-maintenance procedure for support statuses for all nodes as in Table 7.2, assuming that additional justification (SL () ()) has been added to node B. Determine updated values for support statuses, supporting justification, supporting-node sets, antecedent sets, foundation sets, ancestor sets, consequence sets, affected-consequence sets, believed-consequence sets, repercussion sets, and believed-repercussion sets for all nodes,

   b. Same as (a), but the additional justification added to B is (SL (D) (E)),

   c. Same as (a), but the addition justification added to B is(SL (C) ()).

5. For the truth-maintenance system described by the table that follows:

   a. Determine supporting statuses for all nodes,

   b. For supporting statuses of nodes from (a), determine supporting justifications, supporting-node sets, antecedent sets, foundation sets, ancestor sets, consequence sets, affected-consequence sets, believed-consequence sets, repercussion sets, and believed-repercussion sets for all nodes,

c. With the additional justification (SL (AB) ()) for node B, determine updated values for support statuses, supporting justifications, supporting-
   node sets, antecedent sets, foundation sets, ancestor sets, consequence
   sets, affected-consequence sets, believed-consequence sets, repercussion
   sets, and believed-repercussion sets for all nodes.

Node	Justification Set
A	{J1 = (SL () ())}
B	
C	{J2 = (SL () ())}
D	{J3 = (SL (AB) ())}
E	{J4 = (SL () (B))}
F	
G	{J5 = (SL (E) (D)), J6 = (SL (C) ())}.

REFERENCES

Adlassnig, K.-P. (1982). A survey on medical diagnosis and fuzzy subsets. In *Approximate Reasoning in Decision Analysis*, M. M. Gupta, E. Sanchez (eds.), North-Holland, 203–217.

Adlassnig, K.-P. and G. Kolarz (1982). CADIAG-2: Computer-assisted medical diagnosis using fuzzy subsets. In *Approximate Reasoning in Decision Analysis*, M. M. Gupta, E. Sanchez (eds.), North-Holland, 219–247.

Baldwin, J. F. (1979a). A new approach to approximate reasoning using a fuzzy logic. *Fuzzy Sets and Systems* 2, 309–325.

Baldwin, J. F. (1979b). Fuzzy logic and fuzzy reasoning. *Int. J. Man-Machine Studies* 11, 465–480.

Bellman, R. E. and M. Giertz (1973). On the analytic formalism of the theory of fuzzy sets. *Inf. Sci.* 5, 149–157.

Biswas, G. and T. S. Anand (1987). Using the Dempster-Shafer scheme in a diagnostic system shell. *Proc. 3rd Workshop Uncertainty in AI*, July 13–17, Seattle, WA, 98–105.

Boose, J. H. (1989). A survey of knowledge acquisition techniques and tools. *Knowledge Acquisition* 1, 3–37.

Bouchon, B. and R. R. Yager (eds.) (1987). *Uncertainty in Knowledge-Based Systems. Int. Conf. Information Processing and Management of Uncertainty in Knowledge-Based Systems*, June 30–July 4, 1986, Paris, France. *Selected and Extended Contributions. Lecture Notes in Computer Science* 286, Springer-Verlag.

Bouchon, B., L. Saitta, and R. R. Yager (eds.), (1988). *Uncertainty and Intelligent Systems. 2nd Int. Conf. Information Processing and Management of Uncertainty in Knowledge-Based Systems IPMU'88*, July 4–7, 1988, Urbino, Italy. *Lecture Notes in Computer Science* 313, Springer-Verlag.

Buchanan, B. G., D. Barstow, R. Bechtal, J. Benett, W. Clancey, C. Kulikowski, T. Mitchell, and D. A. Waterman (1983). Constructing an expert system. In *Building Expert Systems*, F. Hayes-Roth, D. A. Waterman, D. B. Lenat (eds.), Addison-Wesley, 127–167.

Buchanan, B. G. and R. D. Duda (1983). Principles of rule-based expert systems. *Advances in Computers* 22, 163–216.

Buchanan, B. G. and E. A. Feigenbaum (1978). DENDRAL and META-DENDRAL: Their applications dimension. *AI* 11, 5–24.

Bundy, A. (1985). Incidence calculus: a mechanism for probabilistic reasoning. *J. Automated Reasoning* 1, 263–283.

Bundy, A. (1986). Correctness criteria of some algorithms for uncertain reasoning using incidence calculus. *J. Automated Reasoning* 2, 109–126.

Carbonell, J. G. and A. Collins (1973). Natural semantics in artificial intelligence. *Proc. 3rd IJCAI*, Stanford, CA, 344–351.

Cheeseman, P. (1983). A method of computing generalized Bayesian probability values for expert systems. *Proc. 8th IJCAI*, Aug. 8–12, 1983, Karlsruhe, W. Germany, 198–202.

Cheeseman, P. (1985). In defense of probability. *Proc 9th IJCAI, Aug. 18–23, 1985, Los Angeles, CA*, 1002–1009.

Cheeseman, P. (1986). Probabilistic vs. fuzzy reasoning. In *Uncertainty in Artificial Intelligence*, L. N. Kanal, J. F. Lemmer (eds.), North Holland, 85–102.

Cohen, P. R. and M. R. Grinberg (1983). A framework for heuristic reasoning about uncertainty. *Proc. 8-h IJCAI*, Aug. 8–12, 1983, Karlsruhe, W. Germany, 355–357.

Collins, A. (1978). Fragments of a theory of human plausible reasoning. In *Theoretical Issues in Natural Language Processing*, D. Waltz (Ed.), University of Illinois.

Collins, A. and R. Michalski (1986). The logic of plausible reasoning: A core theory. Department of Computer Science, University of Illinois at Urbana-Champaign, UIUCDCS-F-86-951.

D'Ambrosio, B. (1987). Truth maintenance with numeric certainty estimates. *Proc. 3rd Conf. on AI Applications*, Feb. 23–27, 1987, Kissimmee, FL, 244–249.

Dean, J. S. and J. W. Grzymala-Busse (1988). An overview of the learning from examples module LEM1. Department of Computer Science, University of Kansas, TR-88-2.

de Kleer, J. (1986). An assumption-based truth maintenance system. *AI* 28, 127–162.

Deliyanni, A. and R. A. Kowalski (1979). Logic and semantic networks. *Com. ACM* 22, 184–192.

di Nola, A., S. Sessa, W. Pedrycz, and E. Sanchez (1989). *Fuzzy Relation Equations and Their Applications to Knowledge Engineering*. Kluwer.

Dietterich, T. G. and R. S. Michalski (1983). A comparative review of selected methods for learning from examples. In *Machine Learning. An Artificial Intelligence Approach*, R. S. Michalski, J. G. Carbonell, T. M. Mitchell (eds.), Morgan Kaufmann, 41–81.

Dontas, K. and M. Zemankova (1988). APPLAUS: An experimental plausible reasoning system. *Proc. 3rd Int. Symp. Methodologies for Intelligent Systems*, Oct. 12–15, 1988, Turin, Italy, 29–39.

Doyle, J. (1979). A truth maintenance system. *AI* 12, 231–272.

Doyle, J. (1985). Reasoned assumptions and Pareto optimality. *Proc. 9th IJCAI*, Aug. 18–23, 1985, Los Angeles, CA, 87–90.

Dubois, D. and H. Prade (1980). *Fuzzy Sets and Systems: Theory and Applications*. Academic Press.

Dubois, D. and H. Prade (1985). Combination and propagation of uncertainty with belief functions. A reexamination. *Proc. 9th IJCAI 85*, Aug. 18–23, 1985, Los Angeles, CA, 111–113.

Dubois, D. and H. Prade (1988a). An introduction to possibilistic and fuzzy logics. In *Non-Standard logics for Automated Reasoning*, P. Smets, E. H. Mamdani, D. Dubois, H. Prade (eds.), Academic Press, 287–315.

Dubois, D. and H. Prade (1988b). Default reasoning and possibility theory. *AI* 35, 243–257.

Dubois, D. and H. Prade (1988c). *Possibility Theory. An Approach to Computerized Processing of Uncertainty*. Plenum Press.

Duda, R., P. Hart, and N. Nilsson (1976). Subjective Bayesian methods for rule-based inference systems. In *Proc. 1976 Nat. Computer Conf., AFIPS 45*, 1976, 1075–1082.

Duda, R., J. Gaschnig, and P. Hart (1979). Model design in the Prospector consultant system for mineral exploration. In *Expert Systems Microelectronic Age*, D. Michie (ed.), Edinburgh U. Press, 153–167.

Etherington, D. W. (1987). Relating default logic and circumscription. *Proc. 10th IJCAI 87*, Aug. 23–28, 1987, Milano, Italy, 489–494.

Forsyth, D. E. and B. G. Buchanan (1989). Knowledge acquisition for expert systems: Some pitfalls and suggestions. *IEEE Trans. on Systems, Man, and Cybernetics* 19, 435–442.

Forgy, C. L. (1982). Rete: A fast algorithm for the many pattern/ many object pattern match problem. *AI* 19, 17–37.

Frost, R. (1986). *Introduction to Knowledge Base Systems*. Macmillan Publ. Co.

Giles, R. (1976). Lukasiewicz logic and fuzzy theory. *Int. J. Man-Machine Studies* 8, 313–327.

Ginsberg, M. L. (1984). Non-monotonic reasoning using Dempster's rule. *Proc. Nat. Conf. AAAI-84*, Aug. 6–10, 1984, Austin, TX, 126–129.

Ginsberg, M. L. (1985). Does probability have a place in non-monotonic reasoning? *Proc. 9th IJCAI 85*, Aug. 18–23, 1987, Los Angeles, CA, 107–110.

Goodman, I. R. and H. T. Nguyen (1985). *Uncertainty Models for Knowledge-Based Systems*. North-Holland.

Gordon, J. and E. H. Shortliffe (1984). The Dempster-Shafer theory of evidence. In *Rule Based Expert Systems. The MYCIN Experiments of the Stanford Heuristic Programming Project*, B. G. Buchanan, E. F. Shortliffe (eds.), Addison-Wesley, 272–292.

Graham, I. and P. L. Jones (1988). *Expert Systems: Knowledge, Uncertainty and Design*. Chapman and Hall.

Grosof, B. (1984). Default reasoning as circumscription. In *Workshop on Non-Monotonic Reasoning*, Oct. 17–19, New Paltz, NY, 115–124.

Grosof, B. (1986). Evidential confirmation as transformed probability. On the duality of priors and updates. In *Uncertainty in Artificial Intelligence*, L. N. Kanal, J. F. Lemmer (eds.), North Holland, 153–166.

Grzymala-Busse, J. W. (1988). Knowledge acquisition under uncertainty—a rough set approach. *J. Intelligent Robotic Systems* 1, 3–16.

Grzymala-Busse, J. W. and D. J. Sikora (1988). LERS1—a system for learning from examples based on rough sets. Department of Computer Science, University of Kansas, TR-88-5.

Gupta, M. M. and T. Yamakawa (eds.) (1988a). *Fuzzy Logic in Knowledge-Based Systems, Decision and Control.* North-Holland.

Gupta, M. M. and T. Yamakawa (eds.), (1988b). *Fuzzy Computing: Theory, Hardware, and Applications.* North-Holland.

Heckerman, D. (1986). Probabilistic interpretations for MYCIN's certainty factors. In *Uncertainty in Artificial Intelligence*, L. N. Kanal, J. F. Lemmer (eds.), North Holland, 167–196.

Hunter, D. (1986). Uncertain reasoning using maximum entropy inference. In *Uncertainty in Artificial Intelligence*, L. N. Kanal, J. F. Lemmer (eds.), North Holland, 203–209.

Imielinski, T. (1985). Results on translating defaults to circumscription. *Proc 9th IJCAI 85*, Aug. 18–23, 1985, Los Angeles, CA, 114–120.

Ishizuka, M., K. S. Fu, and J. T. P. Yao (1982). A rule-based inference with fuzzy set for structural damage assessment. In *Approximate Reasoning in Decision Analysis*, M. M. Gupta, E. Sanchez (eds.), North-Holland, 261–268.

Jackson, P. (1986). *Introduction to Expert Systems.* Addison-Wesley.

Johnson, R. W. (1986). Independence and Bayesian updating methods. *AI 29*, 217–222.

Kanal, L. N. and J. F. Lemmer (eds.) (1986). *Uncertainty in Artificial Intelligence.* North-Holland.

Konolige, K. (1986). *A Deduction Model of Belief.* Pitman/ Morgan Kaufmann.

Konolige, K. G. (1987). On the relation between default theories and autoepistemic logic. *Proc. 10th IJCAI 87*, Aug. 23–28, 1987, Milano, Italy, 394–401.

Kyburg, H. E. (1986). Representing knowledge and evidence for decision. In *Uncertainty in Knowledge-Based Systems. Int. Conf. Information Processing and Management of Uncertainty in Knowledge-Based Systems*, June 30–July 4, 1986, Paris, France, 30–40.

Kyburg, H. E. (1987a). Objective probabilities. *Proc. 10th IJCAI 87*, Aug. 23–28, 1987, Milano, Italy, 902–904.

Kyburg, H. E., Jr. (1987b). Bayesian and non-Bayesian evidential updating. *AI 31*, 271–293.

Laskey, K. B. and P. E. Lehner (1988). Belief maintenance: An integrated approach to uncertainty management. *Proc. AAAI-88, 7th Nat. Conf. AI*, Aug. 21–28, 1988, St. Paul, MN, 210–214.

Lee, N. S., Y. L. Grize, and K. Dehnad (1987). Quantitative models for reasoning under uncertainty in knowledge-based expert systems. *Int. J. Intelligent Systems* 2, 15–38.

Lemmer, J. F. and L. N. Kanal (eds.), (1988). *Uncertainty in Artificial Intelligence 2.* North-Holland.

Lemmer, J. F. (1986). Confidence factors, empiricism and the Dempster-Shafer theory of evidence. In *Uncertainty in Artificial Intelligence*, L. N. Kanal, J. F. Lemmer (eds.), North Holland, 117–125.

Liu, X. and A. Gammerman (1987). On the validity and applicability of the IFERNO system. In *Research and Development in Expert Systems III, Proc. of Expert Systems'86*, Dec. 15–18, 1986, Brighton, Great Britain, 47–56.

Lowrance, J. D., T. D. Garvey, and T. M. Strat (1986). A framework for evidential-reasoning systems. *Proc. 5th Nat. Conf. AI, AAAI-86*, Aug. 11–15, 1986, Philadelphia, PA, 896–903, 1986.

Lukaszewicz, W. (1984). Considerations on default logic. In *Workshop on Non-Monotonic Reasoning*, Oct. 17–19, New Paltz, NY, 165–193.

Mamdani, A., J. Efstathiou, and D. Pang (1985). Inference under uncertainty. *Expert Systems 85, Proc. 5th Conf. British Comp. Soc., Specialist Group on Expert Systems*, Dec. 17–19, 1985, University of Warwick, Great Britain, 181 - 190.

Marciszewski, W. (ed.) (1981). *Dictionary of Logic as Applied in the Study of Language. Concepts, Methods, Theories.* Martinus Nijhoff.

McCarthy, J. (1977). Epistemological problems of artificial intelligence. *Proc. 5th IJCAI*, Cambridge, MA, 1038–1044.

McCarthy, J. (1980). Circumscription—a form of non-monotonic reasoning. *AI* 13, 27–39.

McCarthy, J. (1984). Applications of circumscription to formalize common sense knowledge. In *Workshop on Non-Monotonic Reasoning*, Oct. 17–19, New Paltz, NY, 295–324.

McDermott, D. (1982). Nonmonotonic logic II: Nonmonotonic modal theories. *J. ACM* 29, 33–57.

McDermott, D. and J. Doyle (1980). Non-monotonic logic I. *AI* 13, 41–72.

McGraw, K. L. and K. Harbison-Briggs (1989). *Knowledge Acquisition: Principles and Guidelines.* Prentice Hall.

Mellouli, K., G. Shafer, and P. P. Shenoy (1986). Qualitative Markov networks. In *Uncertainty in Knowledge-Based Systems. Int. Conf. Information Processing and Management of Uncertainty in Knowledge-Based Systems*, June 30–July 4, 1986, Paris, France. *Selected and Extended Contributions.Lecture Notes in Computer Science* 286, Springer Verlag, 69–74.

Merkhofer, M. W. (1987). Quantifying judgmental uncertainty: Methodology, experiences, and insights. *IEE Trans. on Systems, Man, and Cybernetics* 17, 741–752.

Michalski, R. S. (1975). Variable-valued logic and its applications to pattern recognition and machine learning. In *Computer Science and Multiple-Valued Logic Theory and Applications*, D.C. Rine (ed.), North-Holland, 506–534.

Michalski, R. S. (1983). A theory and methodology of inductive learning. In *Machine Learning. An Artificial Intelligence Approach*, R. S. Michalski, J. G. Carbonell, T. M. Mitchell (eds.), Morgan Kaufmann, 83–134.

Michalski, R. S. and R. L. Chilausky (1980). Knowledge acquisition by encoding expert rules versus computer induction from examples: A case study involving soybean pathology. *Int. J. Man-Machine Studies* 12, 63–87.

Michalski, R. S. and P. H. Winston (1986). Variable precision logic. *AI* 29, 121–146.

Moore, R. C. (1984). Possible-world semantics for autoepistemic logic. In *Workshop on Non-Monotonic Reasoning*, Oct. 17–19, New Paltz, NY, 344–354.

Moore, R. C. (1985). Semantical considerations on nonmonotonic logic. *AI* 25, 75–94.

Mrozek, A. (1987). Rough sets and some aspects of expert systems realization. *Proc. 7th Int.. Workshop on Expert Systems and their Applications*, May 13–15, 1987, Avignon, France, 597–611.

Nguyen, T. A., W. A. Perkins, T. J. Laffey, and D. Pecora (1987). Knowledge base verification. *AI Magazine* 8, 69–75.

Nilsson, N. J. (1986). Probabilistic logic. *AI* 28, 71–87.

Pawlak, Z. (1982). Rough sets. *Int. J. Comp. Inf. Sci.* 11, 344–356.

Pawlak, Z. (1984). Rough classification. *Int. J. Man-Machine Studies* 20, 469–483.

Pearl, J. (1985). How to do with probability what people say you can't. *2nd Conf. AI Applic., The Engineering of Knowledge-Based Systems*, Dec. 11–133, 1985, Miami Beach, FL, 6–12.

Pearl, J. (1986a). On evidential reasoning in a hierarchy of hypotheses. *AI* 28, 9–15.

Pearl, J. (1986b). Fusion, propagation, and structuring in belief networks. *AI* 29, 241–288.

Pearl, J. (1988). *Probabilistic Reasoning in Intelligent Systems: Networks of Plausible Inference*. Morgan Kaufmann.

Pednault, E. P. D., S. W. Zucker, and L. V. Muresan (1981). On the independence assumption underlying subjective Bayesian updating. *AI* 16, 213–22.

Perez, A. and R. Jirousek (1985). Constructing an intensional expert system (INES). In *Medical Decision Making*. Elsevier.

Quinlan, J. R. (1983a). IFERNO: A cautious approach to uncertain inference. *The Computer J.* 26, 350–390.

Quinlan, J. R. (1983b). Learning efficient classification procedures and their application to chess end games. In *Machine Learning. An Artificial Intelligence Approach*, R. S. Michalski, J. G. Carbonell, T. M. Mitchell (eds.), Morgan Kaufmann, 463–482.

Rauch-Hindin, W. B. (1986). *Artificial Intelligence in Business, Science, and Industry. Vol I—Fundamentals*. Prentice-Hall.

Reichgelt, H. and F. van Harmelen (1985). Relevant criteria for choosing an inference engine in expert systems. *Expert Systems 85, Proc. 5th Conf. British Comp. Soc., Specialist Group on Expert Systems*, Dec. 15–18, 1986, Brighton, Great Britain, 21–30.

Reiter, R. (1980). A logic for default reasoning. *AI* 13, 81–132.

Reiter, R. and J. de Kleer (1987). Foundations of assumption-based truth mainte-nance systems: Preliminary report. *Proc. 8th IJCAI*, Aug. 8–12, 1983, Karl-sruhe, W. Germany, 183–188.

Rescher, N. (1976). *Plausible reasoning*. Van Gorcum.

Rich, E. (1983). Default reasoning as likelihood reasoning. *Proc. Nat. Conf. AAI-83*, Aug. 22–26, 1983, Washington, D.C., 348–351.

Shafer, G. (1976). *A Mathematical Theory of Evidence*. Princeton University Press.

Shafer, G. (1987). Belief functions and possibility measures. In *The Analysis of Fuzzy Information*, J.E. Bezdek (ed.), CRC Press.

Shafer, G. and R. Logan (1987). Implementing Dempster's rule for hierarchical evidence. *AI* 33, 271–248.

Shafer, G., P. P. Shenoy, and K. Mellouli (1987). Propagating belief functions in qualitative Markov trees. *Int. J. Approximate Reasoning* 1, 349–400.

Shafer, G., P. P. Shenoy, and K. Mellouli (1988). Propagation of belief functions: A distributed approach. In *Uncertainty in Artificial Intelligence 2*, J. F. Lem-mer , L. N. Kanal, (eds.), North Holland, 325–335.

Shenoy, P. P. and G. Shafer (1986). Propagating belief functions with local com-putations. *IEEE Expert* 1, 44–52.

Shoham, Y. (1988). *Reasoning about Change. Time and Causation from the Standpoint of Artificial Intelligence*. The MIT Press.

Shore, J. E. (1986). Relative entropy, probabilistic inference, and AI. In *Uncer-tainty in Artificial Intelligence*, L. N. Kanal, J. F. Lemmer (eds.), North Hol-land, 211–215.

Shortliffe, H. and B. G. Buchanan (1975). A model of inexact reasoning in medicine. *Mathematical Biosciences* 23, 351–379.

Smets, P., E. H. Mamdani, D. Dubois, and H. Prade (eds.), (1988). *Non-Standard Logics for Automated Reasoning*. Academic Press.

Steels, L. (1987). Second generation expert systems. In *Research and Develop-ment in Expert Systems III, Proc. of Expert Systems'86*, Dec. 15–18, 1986, Brighton, Great Britain, 175–183.

Sullivan, M. and P. R. Cohen (1985). An endorsement-based plan recognition program. *Proc. IJCAI 85*, Aug. 18–23, Los Angeles, CA, 1985, 475–479.

Suwa, M., A. C. Scott, and E. F. Shortliffe (1984). Completeness and consistency in a rule-based system. In *Rule Based Expert Systems. The MYCIN Experi-ments of the Stanford Heuristic Programming Project*, B. G. Buchanan, E. F. Shortliffe (eds.), Addison-Wesley, 159–170.

Szolowits, P. and S. G. Pauker (1978). Categorical and probabilistic reasoning in medical diagnosis. *AI* 11, 115–144.

Tarski, A (1933). The concept of truth in formalized languages (in Polish), War-saw 1933, German version in *Studia Philosophica* 1, 1936, 261–405, English version in A. Tarski, *Logic, Semantics, Metamathematics: Papers from 1923 to 1938*, Clarendon, 1956.

Togai, M. and H. Watanabe (1986). Expert system on a chip: An engine for real-time approximate reasoning. *Proc. ACM SIGART Int. Symp. Methodologies for Intelligent Systems*, Oct. 22–24, 1986, Knoxville, TN, 147–154.

Waterman, D. A. (1986). *A Guide to Expert Systems*. Addison-Wesley.

Winston, P. H. (1975). Learning structural descriptions from examples. In *The Psychology of Computer Vision*, P. H. Winston (ed.), McGraw-Hill, 157–209. Based on a Ph.D. thesis, MIT, Cambridge, MA, 1970.

Yager, R. R. (1980). On a general class of fuzzy connectives. *Fuzzy Sets and Systems* 3, 235–242.

Yager, R. R. (1987). Using approximate reasoning to represent default knowledge. *AI* 39, 99–112.

Zadeh, L. A. (1965). Fuzzy sets. *Information and Control* 8, 338–353.

Zadeh, L. A. (1978). PRUF—a meaning representation language for natural languages. *Int. J. Man-Machine Studies* 10, 395–460.

Zadeh, L. A. (1979). A theory of approximate reasoning. *Machine Intelligence* 9, 149–194.

Zadeh, L. A. (1981). Possibility theory and soft data analysis. In *Mathematical Frontiers of the Social and Policy Sciences*, L.Cobb, R. M. Thrall (eds.), Westview, 69–129.

Zadeh, L. A. (1986a). A simple view of the Dempster-Shafer theory of evidence and its implication for the rule of combination. *AI Magazine* 7, 85–90.

Zadeh, L. A. (1986b). Is probability theory sufficient for dealing with uncertainty in AI: A negative view. In *Uncertainty in Artificial Intelligence*, L. N. Kanal, J. F. Lemmer (eds.), North Holland, 103–116.

Zarley, D., Y.-T. Hsia, and G. Shafer (1988). Evidential reasoning using DELIEF. *Proc. AAA-88, 7th Nat. Conf. AI*, Aug. 21–26, 1988, St. Paul, MN, 205–209.

Ziarko, W. and J. Katzberg (1989). Control algorithm acquisition, analysis, and reduction. In *Knowledge-Based System Diagnosis, Supervision and Control*, T. Tzafestas (ed.), Plenum Press, 167–178.

I N D E X

217